SpringerBriefs in Probability and Mathematical Statistics

SpringerBriefs present concise summaries of cutting-edge research and practical applications across a wide spectrum of fields. Featuring compact volumes of 50 to 125 pages, the series covers a range of content from professional to academic. Briefs are characterized by fast, global electronic dissemination, standard publishing contracts, standardized manuscript preparation and formatting guidelines, and expedited production schedules.

Typical topics might include:

- A timely report of state-of-the art techniques
- A bridge between new research results, as published in journal articles, and a contextual literature review
- A snapshot of a hot or emerging topic
- Lecture of seminar notes making a specialist topic accessible for non-specialist readers
- SpringerBriefs in Probability and Mathematical Statistics showcase topics of current relevance in the field of probability and mathematical statistics

Manuscripts presenting new results in a classical field, new field, or an emerging topic, or bridges between new results and already published works, are encouraged. This series is intended for mathematicians and other scientists with interest in probability and mathematical statistics. All volumes published in this series undergo a thorough refereeing process.

The SBPMS series is published under the auspices of the Bernoulli Society for Mathematical Statistics and Probability.

More information about this series at http://www.springer.com/series/14353

Victor M. Panaretos • Yoav Zemel

An Invitation to Statistics in Wasserstein Space

Springer Open

Bernoulli Society
for Mathematical Statistics
and Probability

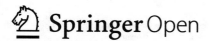

Victor M. Panaretos
Institute of Mathematics
EPFL
Lausanne, Switzerland

Yoav Zemel
Statistical Laboratory
University of Cambridge
Cambridge, UK

ISSN 2365-4333 ISSN 2365-4341 (electronic)
SpringerBriefs in Probability and Mathematical Statistics
ISBN 978-3-030-38437-1 ISBN 978-3-030-38438-8 (eBook)
https://doi.org/10.1007/978-3-030-38438-8

This Springer imprint is published by the registered company Springer Nature Switzerland AG.
The registered company address is: Gewerbestrasse 11, 6330 Cham, Switzerland

To our families

Preface

A Wasserstein distance is a metric between probability distributions μ and ν on a ground space $\mathscr{X}$, induced by the problem of optimal mass transportation or simply *optimal transport*. It reflects the minimal effort that is required in order to reconfigure the mass of μ to produce the mass distribution of ν. The 'effort' corresponds to the total work needed to achieve this reconfiguration, where work equals the amount of mass at the origin times the distance to the prescribed destination of this mass. The distance between origin and destination can be raised to some power other than 1 when defining the notion of work, giving rise to correspondingly different Wasserstein distances. When viewing the space of probability measures on $\mathscr{X}$ as a metric space endowed with a Wasserstein distance, we speak of a *Wassertein Space*.

Mass transportation and the associated Wasserstein metrics/spaces are ubiquitous in mathematics, with a long history that has seen them catalyse core developments in analysis, optimisation, and probability. Beyond their intrinsic mathematical richness, they possess attractive features that make them a versatile tool for the statistician. They frequently appear in the development of statistical theory and inferential methodology, sometimes as a technical tool in asymptotic theory, due to the useful topology they induce and their easy majorisation; and other times as a methodological tool, for example, in structural modelling and goodness-of-fit testing. A more recent trend in statistics is to consider Wasserstein spaces themselves as a sample and/or parameter space and treat inference problems in such spaces. It is this more recent trend that is the topic of this book and is coming to be known as 'statistics in Wasserstein spaces' or 'statistical optimal transport'.

From the theoretical point of view, statistics in Wasserstein spaces represents an emerging topic in mathematical statistics, situated at the interface between functional data analysis (where the data are functions, seen as random elements of an infinite-dimensional Hilbert space) and non-Euclidean statistics (where the data satisfy non-linear constraints, thus lying on non-Euclidean manifolds). Wasserstein spaces provide the natural mathematical formalism to describe data collections that are best modelled as random measures on $\mathbb{R}^d$ (e.g. images and point processes). Such random measures carry the infinite-dimensional traits of functional data, but are intrinsically non-linear due to positivity and integrability restrictions. Indeed, contrarily to functional data, their dominating statistical variation arises through random (non-linear) deformations of an underlying template, rather than the additive (linear) perturbation of an underlying template. This shows optimal transport to be a canonical framework for dealing with problems involving the so-called *phase variation* (also known as registration, multi-reference alignment, or synchronisation problems). This connection is pursued in detail in this book and linked with the so-called problem of optimal multitransport (or optimal multicoupling).

In writing our monograph, we had two aims in mind:

1. To present the key aspects of optimal transportation and Wasserstein spaces (Chaps. 1 and 2) relevant to statistical inference, tailored to the interests and background of the (mathematical) statistician. There are, of course, classic texts comprehensively covering this background.[1] But their choice of topics and style of exposition are usually adapted to the analyst and/or probabilist, with aspects most relevant for statisticians scattered among (much) other material.
2. To make use of the 'Wasserstein background' to present some of the fundamentals of statistical estimation in Wasserstein spaces, and its connection to the problem of phase variation (registration) and optimal multicoupling. In doing so, we highlight connections with classical topics in statistical shape theory, such as Procrustes analysis. On these topics, no book/monograph appears to yet exist.

The book focusses on the *theory* of statistics in Wasserstein spaces. It does not cover the associated computational/numerical aspects. This is partially due to space restrictions, but also due to the fact that a reference entirely dedicated to such issues can be found in the very recent monograph of Peyré and Cuturi [103]. Moreover, since this book is meant to be a rapid introduction for non-specialists, we have made no attempt to give a complete bibliography. We have added some bibliographic remarks at the end of each chapter, but these are in no way meant to be exhaustive. For those seeking reference works, Rachev [106] is an excellent overview of optimal transport up to 1985. Other recent reviews are Bogachev and Kolesnikov [26] and Panaretos and Zemel [101]. The latter review can be thought of as complementary to the present book and surveys some of the applications of optimal transport methods to statistics and probability theory.

[1] E.g. by Rachev and Rüschendorf [107], Villani [124, 125], Ambrosio and Gigli [10], Ambrosio et al. [12], and more recently by Santambrogio [119].

Structure of the Book

The material is organised into five chapters.

- Chapter 1 presents the necessary background in optimal transportation. Starting with Monge's original formulation, it presents Kantorovich's probabilistic relaxation and the associated duality theory. It then focusses on quadratic cost functions (squared normed cost) and gives a more detailed treatment of certain important special cases. Topics of statistical concern such as the regularity of transport maps and their stability under weak convergence of the origin/destination measures are also presented. The chapter concludes with a consideration of more general cost functions and the characterisation of optimal transport plans via cyclical monotonicity.
- Chapter 2 presents the salient features of (ℓ_2-)Wasserstein space starting with topological properties of statistical importance, as well as metric properties such as covering numbers. It continues with geometrical features of the space, reviewing the tangent bundle structure of the space, the characterisation of geodesics, and the log and exponential maps as related to transport maps. Finally, it reviews the relationship between the curvature and the so-called compatibility of transport maps, roughly speaking when can one expect optimal transport maps to form a group.
- Chapter 3 starts to shift attention to issues more statistical and treats the problem of existence, uniqueness, characterisation, and regularity of Fréchet means (barycenters) for collections of measures in Wasserstein space. This is done by means of the so-called *multimarginal transport problem* (a.k.a. optimal multitransport or optimal multicoupling problem). The treatment starts with finite collections of measures, and then considers Fréchet means for (potentially uncountably supported) probability distributions on Wasserstein space and associated measurability concerns.
- Chapter 4 considers the problem of estimation of the Fréchet mean of a probability distribution in Wasserstein space, on the basis of a finite collection of i.i.d. elements from this law observed with 'sampling noise'. It is shown that this problem is inextricably linked to the problem of separation of amplitude and phase variation (a.k.a. registration) of random point patters, where the focus is on estimating the maps yielding the optimal multicoupling rather than the Fréchet mean itself. Nonparametric methodology for solving either problem is reviewed, coupled with associated asymptotic theory and several illustrative examples.
- Chapter 5 focusses on the problem of actually *constructing* the Fréchet mean and/or optimal multicoupling of a collection of measures, which is a necessary step when using the methods of Chap. 4 in practice. It presents the steepest descent algorithm based on the geometrical features reviewed in Chap. 2 and a convergence analysis thereof. Interestingly, it is seen that the algorithm is closely related to Procrustes algorithms in shape theory, and this connection is discussed in depth. Several special cases are reviewed in more detail.

Each chapter comes with some bibliographic notes at the end, giving some background and suggesting further reading. The first two chapters can be used independently as a crash course in optimal transport for statisticians at the MSc or PhD level depending on the audience's background. Proofs that were omitted from the main text due to space limitations have been organised into an online supplement and can be accessed from any online version chapter published on SpringerLink website: https://link.springer.com/book/10.1007/978-3-030-38438-8.

Acknowledgements

We wish to thank three anonymous reviewers for their thoughtful feedback. We are especially indebted to one of them, whose analytical insights were particularly useful. Any errors or omissions are, of course, our own responsibility. Victor M. Panaretos gratefully acknowledges support from a European Research Council Starting Grant. Yoav Zemel was supported by Swiss National Science Foundation Grant # 178220. Finally, we wish to thank Mark Podolskij and Donna Chernyk for their patience and encouragement.

Lausanne, Switzerland — Victor M. Panaretos
Cambridge, UK — Yoav Zemel

Contents

Chapter 1
Optimal Transport

In this preliminary chapter, we introduce the problem of optimal transport, which is the main concept behind Wasserstein spaces. General references on this topic are the books by Rachev and Rüschendorf [107], Villani [124, 125], Ambrosio et al. [12], Ambrosio and Gigli [10], and Santambrogio [119]. This chapter includes only few proofs, when they are simple, informative, or are not easily found in one of the cited references.

1.1 The Monge and the Kantorovich Problems

In 1781, Monge [95] asked the following question: given a pile of sand and a pit of equal volume, how can one optimally transport the sand into the pit? In modern mathematical terms, the problem can be formulated as follows. There is a sand space $\mathscr{X}$, a pit space $\mathscr{Y}$, and a cost function $c : \mathscr{X} \times \mathscr{Y} \to \mathbb{R}$ that encapsulates how costly it is to move a unit of sand at $x \in \mathscr{X}$ to a location $y \in \mathscr{Y}$ in the pit. The sand distribution is represented by a measure μ on $\mathscr{X}$, and the shape of the pit is described by a measure ν on $\mathscr{Y}$. Our decision where to put each unit of sand can be thought of as a function $T : \mathscr{X} \to \mathscr{Y}$, and it incurs a total transport cost of

$$C(T) = \int_{\mathscr{X}} c(x, T(x))\, \mathrm{d}\mu(x).$$

Moreover, one cannot put all the sand at a single point y in the pit; it is not allowed to shrink or expand the sand. The map T must be mass-preserving: for any subset $B \subseteq$

Electronic Supplementary Material The online version of this chapter (https://doi.org/10.1007/978-3-030-38438-8_1) contains supplementary material.

V. M. Panaretos, Y. Zemel, *An Invitation to Statistics in Wasserstein Space*, SpringerBriefs in Probability and Mathematical Statistics,
https://doi.org/10.1007/978-3-030-38438-8_1

$\mathscr{Y}$ representing a region of the pit of volume $\nu(B)$, exactly that same volume of sand must go into B. The amount of sand allocated to B is $\{x \in \mathscr{X} : T(x) \in B\} = T^{-1}(B)$, so the mass preservation requirement is that $\mu(T^{-1}(B)) = \nu(B)$ for all $B \subseteq \mathscr{Y}$. This condition will be denoted by $T\#\mu = \nu$ and in words: ν is the push-forward of μ under T, or T pushes μ forward to ν. To make the discussion mathematically rigorous, we must assume that c and T are measurable maps, and that $\mu(T^{-1}(B)) = \nu(B)$ for all measurable subsets of $\mathscr{Y}$. When the underlying measures are understood from the context, we call T a *transport map*. Specifying $B = \mathscr{Y}$, we see that no such T can exist unless $\mu(\mathscr{X}) = \nu(\mathscr{Y})$; we shall assume that this quantity is finite, and by means of normalisation, that μ and ν are probability measures. In this setting, the Monge problem is to find the optimal transport map, that is, to solve

$$\inf_{T:T\#\mu=\nu} C(T).$$

We assume throughout this book that $\mathscr{X}$ and $\mathscr{Y}$ are complete and separable metric spaces,[1] endowed with their *Borel σ-algebra*, which, we recall, is defined as the smallest σ-algebra containing the open sets. Measures defined on the Borel σ-algebra of $\mathscr{X}$ are called *Borel measures*. Thus, if μ is a Borel measure on $\mathscr{X}$, then $\mu(A)$ is defined for any A that is open, or closed, or a countable union of closed sets, etc., and any continuous map on $\mathscr{X}$ is measurable. Similarly, we endow $\mathscr{Y}$ with its Borel σ-algebra. The product space $\mathscr{X} \times \mathscr{Y}$ is also complete and separable when endowed with its product topology; its Borel σ-algebra is generated by the product σ-algebra of those of $\mathscr{X}$ and $\mathscr{Y}$; thus, any continuous cost function $c : \mathscr{X} \times \mathscr{Y} \to \mathbb{R}$ is measurable. It will henceforth always be assumed, without explicit further notice, that μ and ν are Borel measures on $\mathscr{X}$ and $\mathscr{Y}$, respectively, and that the cost function is continuous and nonnegative.

It is quite natural to assume that the cost is an increasing function of the distance between x and y, such as a power function. More precisely, that $\mathscr{Y} = \mathscr{X}$ is a complete and separable metric space with metric d, and

$$c(x,y) = d^p(x,y), \qquad p \geq 0, \quad x,y \in \mathscr{X}. \tag{1.1}$$

In particular, c is continuous, hence measurable, if $p > 0$. The limit case $p = 0$ yields the discontinuous function $c(x,y) = \mathbf{1}\{x = y\}$, which nevertheless remains measurable because the diagonal $\{(x,x) : x \in \mathscr{X}\}$ is measurable in $\mathscr{X} \times \mathscr{X}$. Particular focus will be put on the quadratic case $p = 2$ (Sect. 1.6) and the linear case $p = 1$ (Sect. 1.8.2).

The problem introduced by Monge [95] is very difficult, mainly because the set of transport maps $\{T : T\#\mu = \nu\}$ is intractable. And, it may very well be empty: this will be the case if μ is a Dirac measure at some $x_0 \in \mathscr{X}$ (meaning that $\mu(A) = 1$ if $x_0 \in A$ and 0 otherwise) but ν is not. Indeed, in that case the set $B = \{T(x_0)\}$ satisfies $\mu(T^{-1}(B)) = 1 > \nu(B)$, so no such T can exist. This also shows that the problem is asymmetric in μ and ν: in the Dirac example, there always exists a map T such that $T\#\nu = \mu$—the constant map $T(x) = x_0$ for all x is the unique such map. A less

[1] But see the bibliographical notes for some literature on more general spaces.

extreme situation occurs in the case of absolutely continuous measures. If μ and ν have densities f and g on $\mathbb{R}^d$ and T is continuously differentiable, then $T\#\mu = \nu$ if and only if for μ-almost all x

$$f(x) = g(T(x))|\det \nabla T(x)|.$$

This is a highly non-linear equation in T, nowadays known as a particular case of a family of partial differential equations called *Monge–Ampère equations*. More than two centuries after the work of Monge, Caffarelli [32] cleverly used the theory of Monge–Ampère equations to show smoothness of transport maps (see Sect. 1.6.4).

As mentioned above, if $\mu = \delta\{x_0\}$ is a Dirac measure and ν is not, then no transport maps from μ to ν can exist, because the mass at x_0 must be sent to a unique point x_0. In 1942, Kantorovich [77] proposed a relaxation of Monge's problem in which mass can be split. In other words, for each point $x \in \mathscr{X}$ one constructs a probability measure μ_x that describes how the mass at x is split among different destinations. If μ_x is a Dirac measure at some y, then all the mass at x is sent to y. The formal mathematical object to represent this idea is a probability measure π on the product space $\mathscr{X} \times \mathscr{Y}$ (which is $\mathscr{X}^2$ in our particular setting). Here $\pi(A \times B)$ is the amount of sand transported from the subset $A \subseteq \mathscr{X}$ into the part of the pit represented by $B \subseteq \mathscr{Y}$. The total mass sent from A is $\pi(A \times \mathscr{Y})$, and the total mass sent into B is $\pi(\mathscr{X} \times B)$. Thus, π is mass-preserving if and only if

$$\begin{aligned} \pi(A \times \mathscr{Y}) &= \mu(A), \qquad A \subseteq \mathscr{X} \quad \text{Borel;} \\ \pi(\mathscr{X} \times B) &= \nu(B), \qquad B \subseteq \mathscr{Y} \quad \text{Borel.} \end{aligned} \tag{1.2}$$

Probability measures satisfying (1.2) will be called *transference plans*, and the set of those will be denoted by $\Pi(\mu, \nu)$. We also say that π is a *coupling* of μ and ν, and that μ and ν are the first and second *marginal distributions*, or simply *marginals*, of π. The total cost associated with $\pi \in \Pi(\mu, \nu)$ is

$$C(\pi) = \int_{\mathscr{X} \times \mathscr{Y}} c(x,y)\, \mathrm{d}\pi(x,y).$$

In our setting of a complete separable metric space $\mathscr{X}$, one can represent π as a collection of probability measures $\{\pi_x\}_{x \in \mathscr{X}}$ on $\mathscr{Y}$, in the sense that for all measurable nonnegative g

$$\int_{\mathscr{X} \times \mathscr{Y}} g(x,y)\, \mathrm{d}\pi(x,y) = \int_{\mathscr{X}} \left[\int_{\mathscr{Y}} g(x,y)\, \mathrm{d}\pi_x(y) \right] \mathrm{d}\mu(x).$$

The collection $\{\pi_x\}$ is that of the *conditional distributions*, and the iteration of integrals is called *disintegration*. For proofs of existence of conditional distributions, one can consult Dudley [47, Section 10.2] or Kallenberg [76, Chapter 5]. Conversely, the measure μ and the collection $\{\pi_x\}$ determine π uniquely by choosing g to be indicator functions. An interpretation of these notions in terms of random variables will be given in Sect. 1.2.

The Kantorovich problem is to find the best transference plan, that is, to solve

$$\inf_{\pi\in\Pi(\mu,\nu)} C(\pi).$$

The Kantorovich problem is a relaxation of the Monge problem, because to each transport map T one can associate a transference plan $\pi = \pi_T$ of the same total cost. To see this, choose the conditional distribution π_x to be a Dirac at $T(x)$. Disintegration then yields

$$C(\pi)=\int_{\mathscr{X}\times\mathscr{Y}} c(x,y)\,\mathrm{d}\pi(x,y)=\int_{\mathscr{X}}\left[\int_{\mathscr{Y}} c(x,y)\,\mathrm{d}\pi_x(y)\right]\,\mathrm{d}\mu(x)=\int_{\mathscr{X}} c(x,T(x))\,\mathrm{d}\mu(x)=C(T).$$

This choice of π satisfies (1.2) because $\pi(A\times B) = \mu(A\cap T^{-1}(B))$ and $\nu(B) = \mu(T^{-1}(B))$ for all Borel $A\subseteq\mathscr{X}$ and $B\subseteq\mathscr{Y}$.

Compared to the Monge problem, the relaxed problem has considerable advantages. Firstly, the set of transference plans is never empty: it always contains the product measure $\mu\otimes\nu$ defined by $[\mu\otimes\nu](A) = \mu(A)\nu(B)$. Secondly, both the objective function $C(\pi)$ and the constraints (1.2) are linear in π, so the problem can be seen as infinite-dimensional linear programming. To be precise, we need to endow the space of measures with a linear structure, and this is done in the standard way: define the space $M(\mathscr{X})$ of all finite signed Borel measures on $\mathscr{X}$. This is a vector space with $(\mu_1+\alpha\mu_2)(A)=\mu_1(A)+\alpha\mu_2(A)$ for $\alpha\in\mathbb{R}$, $\mu_1,\mu_2\in M(\mathscr{X})$ and $A\subseteq\mathscr{X}$ Borel. The set of probability measures on $\mathscr{X}$ is denoted by $P(\mathscr{X})$, and is a convex subset of $M(\mathscr{X})$. The set $\Pi(\mu,\nu)$ is then a convex subset of $P(\mathscr{X}\times Y)$, and as $C(\pi)$ is linear in π, the set of minimisers is a convex subset of $\Pi(\mu,\nu)$. Thirdly, there is a natural symmetry between $\Pi(\mu,\nu)$ and $\Pi(\nu,\mu)$. If π belongs to the former and we define $\tilde{\pi}(B\times A) = \pi(A\times B)$, then $\tilde{\pi}\in\Pi(\nu,\mu)$. If we set $\tilde{c}(y,x) = c(x,y)$, then

$$C(\pi) = \int_{\mathscr{X}\times\mathscr{Y}} c(x,y)\,\mathrm{d}\pi(x,y) = \int_{\mathscr{Y}\times\mathscr{X}} \tilde{c}(y,x)\,\mathrm{d}\tilde{\pi}(y,x) = \tilde{C}(\tilde{\pi}).$$

In particular, when $\mathscr{X} = \mathscr{Y}$ and $c=\tilde{c}$ is symmetric (as in (1.1)),

$$\inf_{\pi\in\Pi(\mu,\nu)} C(\pi) = \inf_{\tilde{\pi}\in\Pi(\nu,\mu)} \tilde{C}(\tilde{\pi}),$$

and $\pi\in\Pi(\mu,\nu)$ is optimal if and only if its natural counterpart $\tilde{\pi}$ is optimal in $\Pi(\nu,\mu)$. This symmetry will be fundamental in the definition of the Wasserstein distances in Chap. 2.

Perhaps most importantly, a minimiser for the Kantorovich problem exists under weak conditions. In order to show this, we first recall some definitions. Let $C_b(\mathscr{X})$ be the space of real-valued, continuous bounded functions on $\mathscr{X}$. A sequence of probability measures $\{\mu_n\}\in M(\mathscr{X})$ is said to converge *weakly*[2] to $\mu\in M(\mathscr{X})$ if for all $f\in C_b(\mathscr{X})$, $\int f\,\mathrm{d}\mu_n\to\int f\,\mathrm{d}\mu$. To avoid confusion with other types of convergence, we will usually write $\mu_n\to\mu$ weakly; in the rare cases where a symbol

[2] Weak convergence is sometimes called narrow convergence, weak* convergence, or convergence in distribution.

is needed we shall use the notation $\mu_n \xrightarrow{w} \mu$. Of course, if $\mu_n \to \mu$ weakly and $\mu_n \in P(\mathscr{X})$, then μ must be in $P(\mathscr{X})$ too (this is seen by taking $f \equiv 1$ and by observing that $\int f \, d\mu \geq 0$ if $f \geq 0$).

A collection of probability measures $\mathscr{K}$ is *tight* if for all $\varepsilon > 0$ there exists a compact set K such that $\inf_{\mu \in \mathscr{K}} \mu(K) > 1 - \varepsilon$. If $\mathscr{K}$ is represented by a sequence $\{\mu_n\}$, then Prokhorov's theorem (Billingsley [24, Theorem 5.1]) states that a subsequence of $\{\mu_n\}$ must converge weakly to some probability measure μ.

We are now ready to show that the Kantorovich problem admits a solution when c is continuous and nonnegative and $\mathscr{X}$ and $\mathscr{Y}$ are complete separable metric spaces. Let $\{\pi_n\}$ be a minimising sequence for C. Then, according to [24, Theorem 1.3], μ and ν must be tight. If K_1 and K_2 are compact with $\mu(K_1), \nu(K_2) > 1 - \varepsilon$, then $K_1 \times K_2$ is compact and for all $\pi \in \Pi(\mu, \nu)$, $\pi(K_1 \times K_2) > 1 - 2\varepsilon$. It follows that the entire collection $\Pi(\mu, \nu)$ is tight, and by Prokhorov's theorem π_n has a weak limit π after extraction of a subsequence. For any integer K, $c_K(x,y) = \min(c(x,y), K)$ is a continuous bounded function, and

$$C(\pi_n) = \int c(x,y) \, d\pi_n(x,y) \geq \int c_K(x,y) \, d\pi_n(x,y) \to \int c_K(x,y) \, d\pi(x,y), \qquad n \to \infty.$$

By the monotone convergence theorem

$$\liminf_{n\to\infty} C(\pi_n) \geq \lim_{K\to\infty} \int c_K(x,y) \, d\pi(x,y) = C(\pi) \qquad \text{if } \pi_n \to \pi \text{ weakly.} \tag{1.3}$$

Since $\{\pi_n\}$ was chosen as a minimising sequence for C, π must be a minimiser, and existence is established.

As we have seen, the Kantorovich problem is a relaxation of the Monge problem, in the sense that

$$\inf_{T:T\#\mu=\nu} C(T) = \inf_{\pi_T:T\#\mu=\nu} C(\pi) \geq \inf_{\pi\in\Pi(\mu,\nu)} C(\pi) = C(\pi^*),$$

for some optimal π^*. If $\pi^* = \pi_T$ for some transport map T, then we say that the solution is induced from a transport map. This will happen in two different and important cases that are discussed in Sects. 1.3 and 1.6.1.

A remark about terminology is in order. Many authors talk about the *Monge–Kantorovich problem* or the *optimal transport(ation) problem*. More often than not, they refer to what we call here the Kantorovich problem. When one of the scenarios presented in Sects. 1.3 and 1.6.1 is considered, this does not result in ambiguity.

1.2 Probabilistic Interpretation

The preceding section was an analytic presentation of the Monge and the Kantorovich problems. It is illuminating, however, to also recast things in probabilistic terms, and this is the topic of this section.

A *random element* on a complete separable metric space (or any topological space) $\mathscr{X}$ is simply a measurable function X from some (generic) probability space $(\Omega, \mathscr{F}, \mathbb{P})$ to $\mathscr{X}$ (with its Borel σ-algebra). The *probability law* (or *probability distribution*) is the probability measure $\mu_X = X\#\mathbb{P}$ defined on the space $\mathscr{X}$; this is the Borel measure satisfying $\mu_X(A) = \mathbb{P}(X \in A)$ for all Borel sets A.

Suppose that one is given two random elements X and Y taking values in $\mathscr{X}$ and $\mathscr{Y}$, respectively, and a cost function $c : \mathscr{X} \times \mathscr{Y} \to \mathbb{R}$. The Monge problem is to find a measurable function T such that $T(X)$ has the same distribution as Y, and such that the expectation

$$C(T) = \int_{\mathscr{X}} c(x, T(x))\, \mathrm{d}\mu(x) = \int_{\Omega} c[X(\omega), T(X(\omega))]\, \mathrm{d}\mathbb{P}(\omega) = \mathbb{E}c(X, T(X))$$

is minimised.

The Kantorovich problem is to find a joint distribution for the pair (X, Y) whose marginals are the original distributions of X and Y, respectively, and such that the probability law $\pi = (X, Y)\#\mathbb{P}$ minimises the expectation

$$C(\pi) = \int_{\mathscr{X} \times \mathscr{Y}} c(x, y)\, \mathrm{d}\pi(x, y) = \int_{\Omega} c[X(\omega), Y(\omega))]\, \mathrm{d}\mathbb{P}(\omega) = \mathbb{E}_\pi c(X, Y).$$

Any such joint distribution is called a coupling of X and Y. Of course, $(X, T(X))$ is a coupling when $T(X)$ has the same distribution as Y. The measures π_x in the previous section are then interpreted as the conditional distribution of Y given $X = x$.

Consider now the important case where $\mathscr{X} = \mathscr{Y} = \mathbb{R}^d$, $c(x, y) = \|x - y\|^2$, and X and Y are square integrable random vectors ($\mathbb{E}\|X\|^2 + \mathbb{E}\|Y\|^2 < \infty$). Let A and B be the covariance matrices of X and Y, respectively, and notice that the covariance matrix of a coupling π must have the form $C = \begin{pmatrix} A & V \\ V^t & B \end{pmatrix}$ for a $d \times d$ matrix V. The covariance matrix of the difference $X - Y$ is

$$\begin{pmatrix} I_d & -I_d \end{pmatrix} \begin{pmatrix} A & V \\ V^t & B \end{pmatrix} \begin{pmatrix} I_d \\ -I_d \end{pmatrix} = A + B - V^t - V$$

so that

$$\mathbb{E}_\pi c(X, Y) = \mathbb{E}_\pi \|X - Y\|^2 = \|\mathbb{E}X - \mathbb{E}Y\|^2 + \mathrm{tr}_\pi [A + B - V^t - V].$$

Since only V depends on the coupling π, the problem is equivalent to that of maximising the trace of V, the cross-covariance matrix between X and Y. This must be done subject to the constraint that a coupling π with covariance matrix C exists; in particular, C has to be positive semidefinite.

1.3 The Discrete Uniform Case

There is a special case in which the Monge–Kantorovich problem reduces to a finite combinatorial problem. Although it may seem at first hand as an oversimplification of the original problem, it is of importance in practice because arbitrary measures can be approximated by discrete measures by means of the strong law of large numbers. Moreover, the discrete case is important in theory as well, as a motivating example for the Kantorovich duality (Sect. 1.4) and the property of cyclical monotonicity (Sect. 1.7).

Suppose that μ and ν are each uniform on n distinct points:

$$\mu = \frac{1}{n}(\delta\{x_1\} + \cdots + \delta\{x_n\}), \qquad \nu = \frac{1}{n}(\delta\{y_1\} + \cdots + \delta\{y_n\}).$$

The only relevant costs are $c_{ij} = c(x_i, y_j)$, the collection of which can be represented by an $n \times n$ matrix $\mathbf{C}$. Transport maps T are associated with *permutations* in S_n, the set of all bijective functions from $\{1, \ldots, n\}$ to itself: given $\sigma \in S_n$, a transport map can be constructed by defining $T(x_i) = y_{\sigma(i)}$. If σ is not a permutation, then T will not be a transport map from μ to ν. Transference plans π are equivalent to $n \times n$ matrices M with coordinates $M_{ij} = \pi(\{(x_i, y_j)\}) = M_{ij}$; this is the amount of mass sent from x_i to y_j. In order for π to a be a transference plan, it must be that $\sum_j M_{ij} = 1/n$ for all i and $\sum_i M_{ij} = 1/n$ for all j, and in addition M must be nonnegative. In other words, the matrix $M' = nM$ belongs to B_n, the set of bistochastic matrices of order n, defined as $n \times n$ matrices M' satisfying

$$\sum_{j=1}^{n} M'_{ij} = 1, \quad i = 1, \ldots, n; \qquad \sum_{i=1}^{n} M'_{ij} = 1, \quad j = 1, \ldots, n; \qquad M'_{ij} \geq 0.$$

The Monge problem is the combinatorial optimisation problem over permutations

$$\inf_{\sigma \in S_n} C(\sigma) = \frac{1}{n} \inf_{\sigma \in S_n} \sum_{i=1}^{n} c_{i,\sigma(i)},$$

and the Kantorovich problem is the linear program

$$\inf_{nM \in B_n} \sum_{i,j=1}^{n} c_{ij} M_{ij} = \inf_{M \in B_n/n} \sum_{i,j=1}^{n} c_{ij} M_{ij} = \inf_{M \in B_n/n} C(M).$$

If σ is a permutation, then one can define $M = M(\sigma)$ by $M_{ij} = 1/n$ if $j = \sigma(i)$ and 0 otherwise. Then $M \in B_n/n$ and $C(M) = C(\sigma)$. Such M (or, more precisely, nM) is called a *permutation matrix*.

The Kantorovich problem is a linear program with n^2 variables and $2n$ constraints. It must have a solution because B_n (hence B_n/n) is a compact (nonempty) set in $\mathbb{R}^{n^2}$ and the objective function is linear in the matrix elements, hence continuous. (This property is independent of the possibly infinite-dimensional spaces $\mathscr{X}$

and $\mathcal{Y}$ in which the points lie.) The Monge problem also admits a solution because S_n is a finite set. To see that the two problems are essentially the same, we need to introduce the following notion. If B is a convex set, then $x \in B$ is an *extremal point* of B if it cannot be written as a convex combination $tz + (1-t)y$ for some distinct points $y, z \in B$. It is well known (Luenberger and Ye [89, Section 2.5]) that there exists an optimal solution that is extremal, so that it becomes relevant to identify the extremal points of B_n. It is fairly clear that each permutation matrix is extremal in B_n; the less obvious converse is known as Birkhoff's theorem, a proof of which can be found, for instance, at the end of the introduction in Villani [124] or (in a different terminology) in Luenberger and Ye [89, Section 6.5]. Thus, we have:

Proposition 1.3.1 (Solution of Discrete Problem) *There exists $\sigma \in S_n$ such that $M(\sigma)$ minimises $C(M)$ over B_n/n. Furthermore, if $\{\sigma_1, \dots, \sigma_k\}$ is the set of optimal permutations, then the set of optimal matrices is the convex hull of $\{M(\sigma_1), \dots, M(\sigma_k)\}$. In particular, if σ is the unique optimal permutation, then $M(\sigma)$ is the unique optimal matrix.*

Thus, in the discrete case, the Monge and the Kantorovich problems coincide. One can of course use the simplex method [89, Chapter 3] to solve the linear program, but there are $n!$ vertices, and there is in principle no guarantee that the simplex method solves the problem efficiently. However, the constraints matrix has a very specific form (it contains only zeroes and ones, and is totally unimodular), so specialised algorithms for this problem exist. One of them is the Hungarian algorithm of Kuhn [85] or its variant of Munkres [96] that has a worst-case computational complexity of at most $O(n^4)$. Another alternative is the class of net flow algorithms described in [89, Chapter 6]. In particular, the algorithm of Edmonds and Karp [50] has a complexity of at most $O(n^3 \log n)$. This monograph does not focus on computational aspects for optimal transport. This is a fascinating and very active area of contemporary research, and readers are directed to Peyré and Cuturi [103].

Remark 1.3.2 *The special case described here could have been more precisely called "the discrete uniform case on the same number of points", as "the discrete case" could refer to any two finitely supported measures μ and ν. In the Monge context, the setup discussed here is the most interesting case, see page 8 in the supplement for more details.*

1.4 Kantorovich Duality

The discrete case of Sect. 1.3 is an example of a linear program and thus enjoys a rich duality theory (Luenberger and Ye [89, Chapter 4]). The general Kantorovich problem is an infinite-dimensional linear program, and under mild assumptions admits similar duality.

1.4.1 Duality in the Discrete Uniform Case

We can represent any matrix M as a vector in $\mathbb{R}^{n^2}$, say $\mathbf{M}$, by enumeration of the elements row by row. If nM is bistochastic, i.e., $M \in B_n/n$, then the $2n$ constraints can be represented in a $(2n) \times n^2$ matrix A. For instance, if $n = 3$, then

$$A = \begin{pmatrix} 1 & 1 & 1 & & & & & & \\ & & & 1 & 1 & 1 & & & \\ & & & & & & 1 & 1 & 1 \\ 1 & & & 1 & & & 1 & & \\ & 1 & & & 1 & & & 1 & \\ & & 1 & & & 1 & & & 1 \end{pmatrix} \in \mathbb{R}^{6\times 9}.$$

For general n, the constraints read $A\mathbf{M} = n^{-1}(1,\dots,1) \in \mathbb{R}^{2n}$ and A takes the form

$$A = \begin{pmatrix} \mathbf{1}_n & & & \\ & \mathbf{1}_n & & \\ & & \ddots & \\ & & & \mathbf{1}_n \\ I_n & I_n & \dots & I_n \end{pmatrix} \in \mathbb{R}^{2n\times n^2}, \qquad \mathbf{1}_n = (1,\dots,1) \in \mathbb{R}^n,$$

with I_n the $n \times n$ identity matrix. Thus, the problem can be written

$$\min_{\mathbf{M}} \mathbf{C}^t\mathbf{M} \qquad \text{subject to} \qquad A\mathbf{M} = \frac{1}{n}(1,\dots,1) \in \mathbb{R}^{2n}; \quad \mathbf{M} \geq 0.$$

The last constraint is to be interpreted coordinate-wise; all the elements of M must be nonnegative. The *dual problem* is constructed by introducing one variable for each row of A, transposing the constraint matrix and interchanging the roles of the objective vector $\mathbf{C}$ and the constraints vector $b = n^{-1}(1,\dots,1)$. Call the new variables $p_1,\dots,p_n$ and $q_1,\dots,q_n$, and notice that each column of A corresponds to exactly one p_i and one q_j, and that the n^2 columns exhaust all possibilities. Hence, the dual problem is

$$\max_{p,q\in\mathbb{R}^n} b^t \binom{p}{q} = \frac{1}{n}\sum_{i=1}^n p_i + \frac{1}{n}\sum_{j=1}^n q_j \qquad \text{subject to} \quad p_i + q_j \leq c_{ij}, \quad i,j = 1,\dots,n. \tag{1.4}$$

In the context of duality, one uses the terminology *primal problem* for the original optimisation problem. *Weak duality* states that if $\mathbf{M}$ and (p,q) satisfy the respective constraints, then

$$b^t \binom{p}{q} = \sum_i p_i \frac{1}{n} + \sum_j q_j \frac{1}{n} = \sum_{i,j} (p_i + q_j) M_{ij} \leq \sum_{i,j} C_{ij} M_{ij} = \mathbf{C}^t\mathbf{M}.$$

In particular, if equality holds, then $\mathbf{M}$ is primal optimal and (p,q) is dual optimal. *Strong duality* is the nontrivial assertion that there exist $\mathbf{M}^*$ and (p^*,q^*) satisfying $\mathbf{C}^t\mathbf{M}^* = b^t\binom{p^*}{q^*}$.

1.4.2 Duality in the General Case

The vectors $\mathbf{C}$ and $\mathbf{M}$ were obtained from the cost function c and the transference plan π as $C_{ij} = c(x_i,y_j)$ and $M_{ij} = \pi(\{(x_i,y_j)\})$. Similarly, we can view the vectors p and q as restrictions of functions $\varphi : \mathscr{X} \to \mathbb{R}$ and $\psi : \mathscr{Y} \to \mathbb{R}$ of the form $p_i = \varphi(x_i)$ and $q_j = \psi(y_j)$. The constraint vector $b = (\mathbf{1}_n, \mathbf{1}_n)$ can be written as $b_i = \mu(\{x_i\})$ and $b_{n+j} = \nu(\{y_j\})$. In this formulation, the constraint $p_i + q_j \le c_{ij}$ writes $(\varphi,\psi) \in \Phi_c$ with

$$\Phi_c = \big\{(\varphi,\psi) \in L_1(\mu)\times L_1(\nu) : \varphi(x)+\psi(y) \le c(x,y) \text{ for all } x,y\big\},$$

and the dual problem (1.4) becomes

$$\sup_{(\varphi,\psi)\in L_1(\mu)\times L_1(\nu)} \left[\int_{\mathscr{X}} \varphi(x)\,\mathrm{d}\mu(x) + \int_{\mathscr{Y}} \psi(y)\,\mathrm{d}\nu(y)\right] \qquad \text{subject to} \qquad (\varphi,\psi)\in\Phi_c.$$

Simple measure theory shows that the set constraints (1.2) defining the transference plans set $\Pi(\mu,\nu)$ are equivalent to functional constraints. For future reference, we state this formally as:

Lemma 1.4.1 (Functional Constraints) *Let μ and ν be probability measures. Then $\pi \in \Pi(\mu,\nu)$ if and only if for all integrable functions $\varphi \in L_1(\mu)$, $\psi \in L_1(\nu)$,*

$$\int_{\mathscr{X}\times\mathscr{Y}} [\varphi(x)+\psi(y)]\,\mathrm{d}\pi(x,y) = \int_{\mathscr{X}} \varphi(x)\,\mathrm{d}\mu(x) + \int_{\mathscr{Y}} \psi(y)\,\mathrm{d}\nu(y).$$

The proof follows from the fact that (1.2) yields the above equality when φ and ψ are indicator functions. One then uses linearity and approximations to deduce the result.

Weak duality follows immediately from Lemma 1.4.1. For if $\pi \in \Pi(\mu,\nu)$ and $(\varphi,\psi)\in\Phi_c$, then

$$\int_{\mathscr{X}} \varphi(x)\,\mathrm{d}\mu(x) + \int_{\mathscr{Y}} \psi(y)\,\mathrm{d}\nu(y) = \int_{\mathscr{X}\times\mathscr{Y}} [\varphi(x)+\psi(y)]\,\mathrm{d}\pi(x,y) \le C(\pi).$$

Strong duality can be stated in the following form:

Theorem 1.4.2 (Kantorovich Duality) *Let μ and ν be probability measures on complete separable metric spaces $\mathscr{X}$ and $\mathscr{Y}$, respectively, and let $c : \mathscr{X}\times\mathscr{Y} \to \mathbb{R}_+$ be a measurable function. Then*

$$\inf_{\pi\in\Pi(\mu,\nu)}\int_{\mathscr{X}\times\mathscr{Y}} c\,\mathrm{d}\pi = \sup_{(\varphi,\psi)\in\Phi_c}\left[\int_{\mathscr{X}}\varphi\,\mathrm{d}\mu + \int_{\mathscr{Y}}\psi\,\mathrm{d}\nu\right].$$

See the Bibliographical Notes for other versions of the duality.

When the cost function is continuous, or more generally, a countable supremum of continuous functions, the infimum is attained (see (1.3)). The existence of maximisers (φ,ψ) is more delicate and requires a finiteness condition, as formulated in Proposition 1.8.1 below.

The next sections are dedicated to more concrete examples that will be used through the rest of the book.

1.5 The One-Dimensional Case

When $\mathscr{X} = \mathscr{Y} = \mathbb{R}$, the Monge–Kantorovich problem has a particularly simple structure, because the class of "nice" transport maps contains at most a single element. Identify $\mu,\nu \in P(\mathbb{R})$ with their cumulative distribution functions F and G defined by

$$F(t) = \mu((-\infty,t]), \qquad G(t) = \nu((-\infty,t]), \qquad t\in\mathbb{R}.$$

Let the cost function be (momentarily) quadratic: $c(x,y) = |x-y|^2/2$. Since for $x_1 \le x_2$, $y_1 \le y_2$

$$c(y_2,x_1) + c(y_1,x_2) - c(y_1,x_1) - c(y_2,x_2) = (x_2-x_1)(y_2-y_1) \ge 0,$$

it seems natural to expect the optimal transport map to be monotonically increasing. It turns out that, on the real line, there is at most one such transport map: if T is increasing and $T\#\mu = \nu$, then for all $t\in\mathbb{R}$

$$G(t) = \nu((-\infty,t]) = \mu((-\infty,T^{-1}(t)]) = F(T^{-1}(t)).$$

If $t = T(x)$, then the above equation reduces to $T(x) = G^{-1}(F(x))$. This formula determines T uniquely, and has an interesting probabilistic interpretation: it is well-known that if X is a random variable with *continuous* distribution function F, then $F(X)$ follows a uniform distribution on $(0,1)$. Conversely, if U follows a uniform distribution, G is any distribution function, and

$$G^{-1}(u) = \inf G^{-1}([u,1]) = \inf\{x\in\mathbb{R} : G(x)\ge u\}, \qquad 0<u<1,$$

is the *quantile function* of X, then the random variable $G^{-1}(U)$ has distribution function G. We say that G^{-1} is the *left-continuous inverse* of G. In terms of push-forward maps, we can write $F\#\mu = \mathrm{Leb}|_{[0,1]}$ and $G^{-1}\#\mathrm{Leb}|_{[0,1]} = \nu$, with Leb standing for Lebesgue measure, and it is restricted to the interval $[0,1]$. Consequently, if F is continuous and G is arbitrary, then $T\#\mu = \nu$; we can view T as pushing μ forward to ν in two steps: firstly, μ is pushed forward to $\mathrm{Leb}|_{[0,1]}$ and secondly, $\mathrm{Leb}|_{[0,1]}$ is pushed forward to ν.

Using the change of variables formula, we see that the total cost of T is

$$C(T) = \int_{\mathbb{R}} |G^{-1}(F(x)) - x|^2 \,\mathrm{d}\mu(x) = \int_0^1 |G^{-1}(u) - F^{-1}(u))|^2 \,\mathrm{d}u.$$

If F is discontinuous, then $F\#\mu$ is not Lebesgue measure, and T is not necessarily defined. But there will exist an optimal transference plan $\pi \in \Pi(\mu,\nu)$ that is monotone in the following sense: there exists a set $\Gamma \subset \mathbb{R}^2$ such that $\pi(\Gamma) = 1$ and whenever $(x_i, y_i) \in \Gamma$,

$$|y_2 - x_1|^2 + |y_1 - x_2|^2 - |y_1 - x_1|^2 - |y_2 - x_2|^2 \geq 0.$$

Thus, mass at x_1 and x_2 can be split if need be, but in a monotone way. For example, if μ puts mass $1/2$ at $x_1 = -1$ and at $x_2 = 1$ and ν is uniform on $[-1,1]$. Then the transference plan spreads the mass of x_1 uniformly on $[-1,0]$, and the mass of x_2 uniformly on $[0,1]$. This is a particular case of the cyclical monotonicity that will be discussed in Sect. 1.7.

Elementary calculations show that the inequality

$$c(y_2,x_1) + c(y_1,x_2) - c(y_1,x_1) - c(y_2,x_2) \geq 0, \qquad x_1 \leq x_2; \quad y_1 \leq y_2$$

holds more generally than the quadratic cost $c(x,y) = |x-y|^2$. Specifically, it suffices that $c(x,y) = h(|x-y|)$ with h convex on $\mathbb{R}_+$.

Since any distribution can be approximated by continuous distributions, in view of the above discussion, the following result from Villani [124, Theorem 2.18] should not be too surprising.

Theorem 1.5.1 (Optimal Transport in $\mathbb{R}$) *Let $\mu,\nu \in P(\mathbb{R})$ with distribution functions F and G, respectively, and let the cost function be of the form $c(x,y) = h(|x-y|)$ with h convex and nonnegative. Then*

$$\inf_{\pi \in \Pi(\mu,\nu)} C(\pi) = \int_0^1 h(G^{-1}(u) - F^{-1}(u))\,\mathrm{d}u.$$

If the infimum is finite and h is strictly convex, then the optimal transference plan is unique. Furthermore, if F is continuous, then the infimum is attained by the transport map $T = G^{-1} \circ F$.

The prototypical choice for h is $h(z) = |z|^p$ with $p > 1$. This result allows in particular a direct evaluation of the Wasserstein distances for measures on the real line (see Chap. 2).

Note that no regularity is needed in order that the optimal transference plan be unique, unlike in higher dimensions (compare Theorem 1.8.2). The structure of solutions in the concave case ($0 < p < 1$) is more complicated, see McCann [94].

When $p = 1$, the cost function is convex but not strictly so, and solutions will not be unique. However, the total cost in Theorem 1.5.1 admits another representation that is often more convenient.

Proposition 1.5.2 (Quantiles and Distribution Functions) *If F and G are distribution functions, then*

$$\int_0^1 |G^{-1}(u) - F^{-1}(u)|\,\mathrm{d}u = \int_{\mathbb{R}} |G(x) - F(x)|\,\mathrm{d}x.$$

The proof is a simple application of Fubini's theorem; see page 13 in the supplement.

Corollary 1.5.3 *If $c(x,y) = |x - y|$, then under the conditions of Theorem 1.5.1*

$$\inf_{\pi \in \Pi(\mu,\nu)} C(\pi) = \int_{\mathbb{R}} |G(x) - F(x)|\,\mathrm{d}x.$$

1.6 Quadratic Cost

This section is devoted to the specific cost function

$$c(x,y) = \frac{\|x - y\|^2}{2}, \qquad x, y \in \mathscr{X},$$

where $\mathscr{X}$ is a separable Hilbert space. This cost is popular in applications, and leads to a lucid and elegant theory. The factor of $1/2$ does not affect the minimising coupling π and leads to cleaner expressions. (It does affect the optimal dual pair, but in an obvious way.)

1.6.1 The Absolutely Continuous Case

We begin with the Euclidean case, where $\mathscr{X} = \mathscr{Y} = (\mathbb{R}^d, \|\cdot\|)$ is endowed with the Euclidean metric, and use the Kantorovich duality to obtain characterisations of optimal maps.

Since the dual objective function to be maximised

$$\int_{\mathbb{R}^d} \varphi\,\mathrm{d}\mu + \int_{\mathbb{R}^d} \psi\,\mathrm{d}\nu$$

is increasing in φ and ψ, one should seek functions that take values as large as possible subject to the constraint $\varphi(x) + \psi(y) \le \|x - y\|^2/2$. Suppose that an oracle tells us that some $\varphi \in L_1(\mu)$ is a good candidate. Then the largest possible ψ satisfying $(\varphi, \psi) \in \Phi_c$ is

$$\psi(y) = \inf_{x\in\mathbb{R}^d}\left[\frac{\|x-y\|^2}{2} - \varphi(x)\right] = \frac{\|y\|^2}{2} + \inf_{x\in\mathbb{R}^d}\left[\frac{\|x\|^2}{2} - \varphi(x) - \langle x,y\rangle\right].$$

In other words,

$$\widetilde{\psi}(y) := \frac{\|y\|^2}{2} - \psi(y) = \sup_{x\in\mathbb{R}^d}\left[\langle x,y\rangle - \widetilde{\varphi}(x)\right], \qquad \tilde{\varphi}(x) = \frac{\|x\|^2}{2} - \varphi(x).$$

As a supremum over affine functions (in y), $\widetilde{\psi}$ enjoys some useful properties. We remind the reader that a function $f:\mathscr{X}\to\mathbb{R}\cup\{\infty\}$ is *convex* if $f(tx+(1-t)y)\le tf(x)+(1-t)f(y)$ for all $x,y\in\mathscr{X}$ and $t\in[0,1]$. It is *lower semicontinuous* if for all $x\in\mathscr{X}$, $f(x)\le\liminf_{y\to x}f(y)$. Affine functions are convex and lower semicontinuous, and it straightforward from the definitions that both convexity and lower semicontinuity are preserved under the supremum operation. Thus, the function $\widetilde{\psi}$ is convex and lower semicontinuous. In particular, it is Borel measurable due to the following characterisation: f is lower semicontinuous if and only if $\{x: f(x)\le\alpha\}$ is a closed set for all $\alpha\in\mathbb{R}$.

From the preceding subsection, we now know that optimal dual functions φ and ψ must take the form of the difference between $\|\cdot\|^2/2$ and a convex function. Given the vast wealth of knowledge on convex functions (Rockafellar [113]), it will be convenient to work with $\widetilde{\varphi}$ and $\widetilde{\psi}$, and to assume that $\widetilde{\psi}=(\widetilde{\varphi})^*$, where

$$f^*(y) = \sup_{x\in\mathbb{R}^d}[\langle x,y\rangle - f(x)], \qquad y\in\mathbb{R}^d$$

is the *Legendre transform* of f ([113, Chapter 26]; [124, Chapter 2]), and is of fundamental importance in convex analysis. Now by symmetry, one can also replace $\widetilde{\varphi}$ by $(\widetilde{\psi})^*=(\widetilde{\varphi})^{**}$, so it is reasonable to expect that an optimal dual pair should take the form $(\|\cdot\|^2/2-\widetilde{\varphi}, \|\cdot\|^2/2-(\widetilde{\varphi})^*)$, with $\widetilde{\varphi}$ convex and lower semicontinuous.

The alternative representation of the dual objective value as

$$\int_{\mathbb{R}^d}\varphi\,\mathrm{d}\mu + \int_{\mathbb{R}^d}\psi\,\mathrm{d}\nu = \frac{1}{2}\int_{\mathbb{R}^d}\|x\|^2\,\mathrm{d}\mu(x) + \frac{1}{2}\int_{\mathbb{R}^d}\|y\|^2\,\mathrm{d}\nu(y) - \int_{\mathbb{R}^d}\widetilde{\varphi}\,\mathrm{d}\mu - \int_{\mathbb{R}^d}\widetilde{\psi}\,\mathrm{d}\nu$$

is valid under the integrability condition

$$\int_{\mathbb{R}^d}\|x\|^2\,\mathrm{d}\mu(x) + \int_{\mathbb{R}^d}\|y\|^2\,\mathrm{d}\nu(y) < \infty$$

that μ and ν have finite second moments. This condition also guarantees that an optimal φ exists, as the conditions of Proposition 1.8.1 are satisfied. An alternative direct proof for the quadratic case can be found in Villani [124, Theorem 2.9].

Suppose that an optimal φ is found. What can we say about optimal transference plans π? According to the duality, a necessary and sufficient condition is that

$$\int_{\mathbb{R}^d\times\mathbb{R}^d}\frac{\|x-y\|^2}{2}\,\mathrm{d}\pi(x,y) = \int_{\mathbb{R}^d}\varphi\,\mathrm{d}\mu + \int_{\mathbb{R}^d}\psi\,\mathrm{d}\nu,$$

where $\psi = \|\cdot\|^2/2 - (\|\cdot\|^2/2 - \varphi)^*$. Equivalently (using Lemma 1.4.1),

$$\int_{\mathbb{R}^d \times \mathbb{R}^d} [\widetilde{\varphi}(x) + (\widetilde{\varphi})^*(y) - \langle x, y\rangle] \, \mathrm{d}\pi(x,y) = 0. \tag{1.5}$$

Since we have $\widetilde{\varphi}(x) + (\widetilde{\varphi})^*(y) \geq \langle x, y\rangle$ everywhere, the integrand is nonnegative. Hence, the integral vanishes if and only if π is concentrated on the set of (x,y) such that $\widetilde{\varphi}(x) + \widetilde{\varphi}^*(y) = \langle x, y\rangle$. By definition of the Legendre transform as a supremum, this happens if and only if the supremum defining $\widetilde{\varphi}^*(y)$ is attained at x; equivalently

$$\widetilde{\varphi}(z) - \widetilde{\varphi}(x) \geq \langle z - x, y\rangle, \qquad z \in \mathscr{X}.$$

This condition is precisely the definition of y being a *subgradient* of $\widetilde{\varphi}$ at x [113, Chapter 23]. When $\widetilde{\varphi}$ is differentiable at x, its unique subgradient is the gradient $y = \nabla\widetilde{\varphi}(x)$ [113, Theorem 25.1]. If we are fortunate and $\widetilde{\varphi}$ is differentiable everywhere, or even μ-almost everywhere, then the optimal transference plan π is unique, and in fact induced from the transport map $\nabla\widetilde{\varphi}$. The problem, of course, is that $\widetilde{\varphi}$ may fail to be differentiable μ-almost surely. This is remedied by assuming some regularity on the source measure μ in order to make sure that *any* convex function be differentiable μ-almost surely, and is done via the following regularity result, which, roughly speaking, states that convex functions are differentiable almost surely. A stronger version is given in Rockafellar [113, Theorem 2.25], with an alternative proof in Alberti and Ambrosio [6, Chapter 2]. One could also combine the local Lipschitz property of convex functions [113, Chapter 10] with Rademacher's theorem (Villani [125, Theorem 10.8]).

Theorem 1.6.1 (Differentiability of Convex Functions) *Let $f : \mathbb{R}^d \to \mathbb{R} \cup \{\infty\}$ be a convex function with domain* $\mathrm{dom} f = \{x \in \mathbb{R}^d : f(x) < \infty\}$ *and let $\mathscr{N}$ be the set of points at which f is not differentiable. Then $\mathscr{N} \cap \overline{\mathrm{dom} f}$ has Lebesgue measure 0.*

Theorem 1.6.1 is usually stated for the interior of $\mathrm{dom} f$, denoted $\mathrm{int}(\mathrm{dom} f)$, rather than the closure. But, since $A = \mathrm{dom} f$ is convex, its boundary has Lebesgue measure zero. To see this assume first that A is bounded. If $\mathrm{int}A$ is empty, then A lies in a lower dimensional subspace [113, Theorem 2.4]. Otherwise, without loss of generality $0 \in \mathrm{int}A$, and then by convexity of A, $\partial A \subseteq (1+\varepsilon)A$ for all $\varepsilon > 0$. When A is unbounded, write it as $\cup_n A \cap [-n,n]^d$.

Another issue that might arise is that optimal φ's might not exist. This is easily dealt with using Proposition 1.8.1. If we assume that μ and ν have finite second moments:

$$\int_{\mathbb{R}^d} \|x\|^2 \, \mathrm{d}\mu(x) < \infty \qquad \text{and} \qquad \int_{\mathbb{R}^d} \|y\|^2 \, \mathrm{d}\nu(y) < \infty,$$

then any transference plan $\pi \in \Pi(\mu, \nu)$ has a finite cost, as is seen from integrating the elementary inequality $\|x - y\|^2 \leq 2\|x\|^2 + 2\|y\|^2$ and using Lemma 1.4.1:

$$C(\pi) \leq \int_{\mathbb{R}^d \times \mathbb{R}^d} [\|x\|^2 + \|y\|^2] \, \mathrm{d}\pi(x,y) = \int_{\mathbb{R}^d} \|x\|^2 \, \mathrm{d}\mu(x) + \int_{\mathbb{R}^d} \|y\|^2 \, \mathrm{d}\nu(y) < \infty.$$

With these tools, we can now prove a fundamental existence and uniqueness result for the Monge–Kantorovich problem. It has been proven independently by several authors, including Brenier [31], Cuesta-Albertos and Matrán [37], Knott and Smith [83], and Rachev and Rüschendorf [117].

Theorem 1.6.2 (Quadratic Cost in Euclidean Spaces) *Let μ and ν be probability measures on $\mathbb{R}^d$ with finite second moments, and suppose that μ is absolutely continuous with respect to Lebesgue measure. Then the solution to the Kantorovich problem is unique, and is induced from a transport map T that equals μ-almost surely the gradient of a convex function ϕ. Furthermore, the pair $(\|x\|^2/2-\phi, \|y\|^2/2-\phi^*)$ is optimal for the dual problem.*

Proof. To alleviate the notation we write ϕ instead of $\widetilde{\varphi}$. By Proposition 1.8.1, there exists an optimal dual pair (φ,ψ) such that $\phi(x) = \|x\|^2/2-\varphi(x)$ is convex and lower semicontinuous, and by the discussion in Sect. 1.1, there exists an optimal π. Since ϕ is μ-integrable, it must be finite almost everywhere, i.e., $\mu(\mathrm{dom}\phi)=1$. By Theorem 1.6.1, if we define $\mathscr{N}$ as the set of nondifferentiability points of ϕ, then $\mathrm{Leb}(\mathscr{N}\cap\mathrm{dom}\phi)=0$; as μ is absolutely continuous, the same holds for μ. (Here Leb denotes Lebesgue measure.)

We conclude that $\mu(\mathrm{int}(\mathrm{dom}\phi)\setminus\mathscr{N})=1$. In other words, ϕ is differentiable μ-almost everywhere, and so for μ-almost any x, there exists a unique y such that $\phi(x)+\phi^*(y)=\langle x,y\rangle$, and $y=\nabla\phi(x)$. This shows that π is unique and induced from the transport map $\nabla\phi(x)$. The gradient $\nabla\phi$ is Borel measurable, since each of its coordinates can be written as $\limsup_{q\to 0, q\in\mathbb{Q}} q^{-1}(\phi(x+qv)-\phi(x))$ for some vector v (the canonical basis of $\mathbb{R}^d$), which is Borel measurable because the limit superior is taken on countably many functions (and ϕ is measurable because it is lower semicontinuous).

1.6.2 Separable Hilbert Spaces

The finite-dimensionality of $\mathbb{R}^d$ in the previous subsection was only used in order to apply Theorem 1.6.1, so one could hope to extend the results to infinite-dimensional separable Hilbert spaces.

Although there is no obvious parallel for Lebesgue measure (i.e., translation invariant) on infinite-dimensional Banach spaces, one can still define absolute continuity via Gaussian measures. Indeed, $\mu\in P(\mathbb{R}^d)$ is absolutely continuous with respect to Lebesgue measure if and only if the following holds: if $\mathscr{N}\subset\mathbb{R}^d$ is such that $\nu(\mathscr{N})=0$ for any nondegenerate Gaussian measure ν, then $\mu(\mathscr{N})=0$. This definition can be extended to any separable Banach space $\mathscr{X}$ via projections, as follows. Let $\mathscr{X}^*$ be the (topological) dual of $\mathscr{X}$, consisting of all real-valued, continuous linear functionals on $\mathscr{X}$.

Definition 1.6.3 (Gaussian Measures) *A probability measure $\mu\in P(\mathscr{X})$ is a nondegenerate Gaussian measure if for any $\ell\in\mathscr{X}^*\setminus\{0\}$, $\ell\#\mu\in P(\mathbb{R})$ is a Gaussian measure with positive variance.*

Definition 1.6.4 (Gaussian Null Sets and Absolutely Continuous Measures) *A subset $\mathcal{N} \subset \mathcal{X}$ is a Gaussian null set if whenever ν is a nondegenerate Gaussian measure, $\nu(\mathcal{N}) = 0$. A probability measure $\mu \in P(\mathcal{X})$ is absolutely continuous if μ vanishes on all Gaussian null sets.*

Clearly, if ν is a nondegenerate Gaussian measure, then it is absolutely continuous.

As explained in Ambrosio et al. [12, Section 6.2], a version of Rademacher's theorem holds in separable Hilbert spaces: a locally Lipschitz function is Gâteaux differentiable except on a Gaussian null set of $\mathcal{X}$. Theorem 1.6.2 (and more generally, Theorem 1.8.2) extend to infinite dimensions; see [12, Theorem 6.2.10].

1.6.3 The Gaussian Case

Apart from the one-dimensional case of Sect. 1.5, there is another special case in which there is a unique *and* explicit solution to the Monge–Kantorovich problem.

Suppose that μ and ν are Gaussian measures on $\mathbb{R}^d$ with zero means and nonsingular covariance matrices A and B. By Theorem 1.6.2, we know that there exists a unique optimal map T such that $T\#\mu = \nu$. Since linear push-forwards of Gaussians are Gaussian, it seems natural to guess that T should be linear, and this is indeed the case.

Since T is a linear map that should be the gradient of a convex function ϕ, it must be that ϕ is quadratic, i.e., $\phi(x) - \phi(0) = \langle x, Qx\rangle$ for $x \in \mathbb{R}^d$ and some matrix Q. The gradient of ϕ at x is $(Q+Q^t)x$ and the Hessian matrix is $Q+Q^t$. Thus, $T = Q+Q^t$ and since ϕ is convex, T must be positive semidefinite.

Viewing T as a matrix leads to the *Riccati equation* $TAT = B$ (since T is symmetric). This is a quadratic equation in T, and so we wish to take square roots in a way that would isolate T. This is done by multiplying the equation from both sides with $A^{1/2}$:

$$[A^{1/2}TA^{1/2}][A^{1/2}TA^{1/2}] = A^{1/2}TATA^{1/2} = A^{1/2}BA^{1/2} = [A^{1/2}B^{1/2}][B^{1/2}A^{1/2}].$$

All matrices in brackets are positive semidefinite. By taking square roots and multiplying with $A^{-1/2}$, we finally find

$$T = A^{-1/2}[A^{1/2}BA^{1/2}]^{1/2}A^{-1/2}.$$

A straightforward calculation shows that $TAT = B$ indeed, and T is positive definite, hence optimal. To calculate the transport cost $C(T)$, observe that $(T-I)\#\mu$ is a centred Gaussian measure with covariance matrix

$$TAT - TA - AT + A = A + B - A^{1/2}[A^{1/2}BA^{1/2}]^{1/2}A^{-1/2} - A^{-1/2}[A^{1/2}BA^{1/2}]^{1/2}A^{1/2}.$$

If $Y \sim \mathcal{N}(0, C)$, then $\mathbb{E}\|Y\|^2$ equals the trace of C, denoted tr$C$. Hence, by properties of the trace,

$$C(T) = \operatorname{tr}\left[A + B - 2(A^{1/2}BA^{1/2})^{1/2}\right]. \tag{1.6}$$

By continuity arguments, (1.6) is the total transport cost between any two Gaussian distributions with zero means, even if A is singular.

If $AB = BA$, the above formulae simplify to

$$T = B^{1/2}A^{-1/2}, \qquad C(T) = \operatorname{tr}\left[A + B - 2A^{1/2}B^{1/2}\right] = \|A^{1/2} - B^{1/2}\|_F^2,$$

with F the Frobenius norm.

If the means of μ and ν are m and n, one simply needs to translate the measures. The optimal map and the total cost are then

$$Tx = n + A^{-1/2}[A^{1/2}BA^{1/2}]^{1/2}A^{-1/2}(x - m);$$

$$C(T) = \|n - m\|^2 + \operatorname{tr}[A + B - 2(A^{1/2}BA^{1/2})^{1/2}].$$

From this, we can deduce a lower bound on the total cost between *any* two measures in $\mathbb{R}^d$ in terms of their second order structure. This is worth mentioning, because such lower bounds are not very common (the Monge–Kantorovich problem is defined by an infimum, and thus typically easier to bound from above).

Proposition 1.6.5 (Lower Bound for Quadratic Cost) *Let $\mu, \nu \in P(\mathbb{R}^d)$ have means m and n and covariance matrices A and B and let π be the optimal map. Then*

$$C(\pi) \geq \|n - m\|^2 + \operatorname{tr}[A + B - 2(A^{1/2}BA^{1/2})^{1/2}].$$

Proof. It will be convenient here to use the probabilistic terminology of Sect. 1.2. Let X and Y be random variables with distributions μ and ν. Any coupling of X and Y will have covariance matrix of the form $C = \begin{pmatrix} A & V \\ V^t & B \end{pmatrix} \in \mathbb{R}^{2d \times 2d}$ for some matrix $V \in \mathbb{R}^{d \times d}$, constrained so that C is positive semidefinite. This gives the lower bound

$$\inf_{\pi \in \Pi(\mu,\nu)} \mathbb{E}_\pi \|X - Y\|^2 = \|m - n\|^2 + \inf_{\pi \in \Pi(\mu,\nu)} \operatorname{tr}_\pi[A + B - 2V] \geq \|m - n\|^2 + \inf_{V: C \geq 0} \operatorname{tr}[A + B - 2V].$$

As we know from the Gaussian case, the last infimum is given by (1.6).

1.6.4 Regularity of the Transport Maps

The optimal transport map T between Gaussian measures on $\mathbb{R}^d$ is linear, so it is of course very smooth (analytic). The densities of Gaussian measures are analytic too, so that T inherits the regularity of μ and ν. Using the formula for T, one can show that a similar phenomenon takes place in the one-dimensional case. Though we do not have a formula for T at our disposal when μ and ν are general absolutely continuous measures on $\mathbb{R}^d$, $d \geq 2$, it turns out that even in that case, T inherits the regularity of μ and ν if some convexity conditions are satisfied.

To guess what kind of results can be hoped for, let us first examine the case $d = 1$. Let F and G denote the distribution functions of μ and ν, respectively. Suppose that G is continuously differentiable and that $G' > 0$ on some open interval (finite or not) I such that $\nu(I) = 1$. Then the inverse function theorem says that G^{-1} is also continuously differentiable. Recall that the *support* of a (Borel) probability measure μ (denoted $\text{supp}\mu$) is the smallest closed set K such that $\mu(K) = 1$. A simple application of the chain rule (see page 19 in the supplement) gives:

Theorem 1.6.6 (Regularity in $\mathbb{R}$) *Let $\mu, \nu \in P(\mathbb{R})$ possess distribution functions F and G of class C^k, $k \geq 1$. Suppose further that* $\text{supp}\nu$ *is an interval I (possibly unbounded) and that $G' > 0$ on the interior of I. Then the optimal map is of class C^k as well. If $F, G \in C^0$ are merely continuous, then so is the optimal map.*

The assumption on the support of ν is important: if μ is Lebesgue measure on $[0, 1]$ and the support of ν is disconnected, then T cannot even be continuous, no matter how smooth ν is.

The argument above cannot be easily extended to measures on $\mathbb{R}^d$, $d \geq 2$, because there is no explicit formula available for the optimal maps. As before, we cannot expect the optimal map to be continuous if the support of ν is disconnected. It turns out that the condition on the support of ν is not connectedness, but rather convexity. This was shown by Caffarelli, who was able to prove ([32] and the references within) that the optimal maps have the same smoothness as the measures. To state the result, we recall the following notation for an open $\Omega \subseteq \mathbb{R}^d$, $k \geq 0$ and $\alpha \in (0, 1]$. We say that $f \in C^{k,\alpha}(\Omega)$ if all the partial derivatives of order k of f are locally α-Hölder on Ω. For example, if $k = 1$, this means that for any $x \in \Omega$ there exists a constant L and an open ball B containing x such that

$$\|\nabla f(z) - \nabla f(y)\| \leq L\|y - z\|^{\alpha}, \qquad y, z \in B.$$

Note that $f \in C^{k+1} \implies f \in C^{k,\beta} \implies f \in C^{k,\alpha} \implies f \in C^k$, for $0 \leq \alpha \leq \beta \leq 1$ so α gives a "fractional" degree of smoothness for f. Moreover, $C^{k,0} = C^k$ and $C^{k,1}$ is quite close to C^{k+1}, since Lipschitz functions are almost surely differentiable.

Theorem 1.6.7 (Regularity of Transport Maps) *Fix open sets $\Omega_1, \Omega_2 \subseteq \mathbb{R}^d$, with Ω_2 convex, and absolutely continuous measures $\mu, \nu \in P(\mathbb{R}^d)$ with finite second moments and bounded, strictly positive densities f, g, respectively, such that $\mu(\Omega_1) = 1 = \nu(\Omega_2)$. Let ϕ be such that $\nabla\phi\#\mu = \nu$.*

1. *If Ω_1 and Ω_2 are bounded and f, g are bounded below, then ϕ is strictly convex and of class $C^{1,\alpha}(\Omega_1)$ for some $\alpha > 0$.*
2. *If $\Omega_1 = \Omega_2 = \mathbb{R}^d$ and $f, g \in C^{0,\alpha}$, then $\phi \in C^{2,\alpha}(\Omega_1)$.*

If in addition $f, g \in C^{k,\alpha}$, then $\phi \in C^{k+2,\alpha}(\Omega_1)$.

In other words, the optimal map $T = \nabla\phi \in C^{k+1,\alpha}(\Omega_1)$ is one derivative smoother than the densities, so has the same smoothness as the measures μ, ν.

Theorem 1.6.7 will be used in two ways in this book. Firstly, it is used to derive criteria for a Karcher mean of a collection of measures to be the Fréchet mean of that collection (Theorem 3.1.15). Secondly, it allows one to obtain very smooth estimates for the transport maps. Indeed, any two measures μ and ν can be approximated by measures satisfying the second condition: one can approximate them by discrete measures using the law of large numbers and then employ a convolution with, e.g., a Gaussian measure (see, for instance, Theorem 2.2.7). It is not obvious that the transport maps between the approximations converge to the transport maps between the original measures, but we will see this to be true in the next section.

1.7 Stability of Solutions Under Weak Convergence

In this section, we discuss the behaviour of the solution to the Monge–Kantorovich problem when the measures μ and ν are replaced by approximations μ_n and ν_n. Since any measure can be approximated by discrete measures *or* by smooth measures, this allows us to benefit from both worlds. On the one hand, approximating μ and ν with discrete measures leads to the finite discrete problem of Sect. 1.3 that can be solved exactly. On the other hand, approximating μ and ν with Gaussian convolutions thereof leads to very smooth measures (at least on $\mathbb{R}^d$), and so the regularity results of the previous section imply that the respective optimal maps will also be smooth. Finally, in applications, one would almost always observe the measures of interest μ and ν with a certain amount of noise, and it is therefore of interest to control the error introduced by the noise. In image analysis, μ can represent an image that has undergone blurring, or some other perturbation (Amit et al. [13]). In other applications, the noise could be due to sampling variation, where instead of μ one observes a discrete measure μ_N obtained from realisations $X_1,\dots,X_N$ of random elements with distribution μ as $\mu_N = N^{-1}\sum_{i=1}^N \delta\{X_i\}$ (see Chap. 4).

In Sect. 1.7.1, we will see that the optimal transference plan π depends continuously on μ and ν. With this result under one's belt, one can then deduce an analogous property for the optimal map T from μ to ν given some regularity of μ, as will be seen in Sect. 1.7.2.

We shall assume throughout this section that $\mu_n \to \mu$ and $\nu_n \to \nu$ weakly, which, we recall, means that $\int_{\mathscr{X}} f\,\mathrm{d}\mu_n \to \int_{\mathscr{X}} f\,\mathrm{d}\mu$ for all continuous bounded $f : \mathscr{X} \to \mathbb{R}$. The following equivalent definitions for weak convergence will be used not only in this section, but elsewhere as well.

Lemma 1.7.1 (Portmanteau) *Let $\mathscr{X}$ be a complete separable metric space and let $\mu, \mu_n \in P(\mathscr{X})$. Then the following are equivalent:*

- *$\mu_n \to \mu$ weakly;*
- *$F_n(x) \to F(x)$ for any continuity point x of F. Here $\mathscr{X} = \mathbb{R}^d$, F_n is the distribution function of μ_n and F is that of μ;*
- *for any open $G \subseteq \mathscr{X}$, $\liminf \mu_n(G) \geq \mu(G)$;*
- *for any closed $F \subseteq \mathscr{X}$, $\limsup \mu_n(F) \leq \mu(F)$;*
- *$\int h\,\mathrm{d}\mu_n \to \int h\,\mathrm{d}\mu$ for any bounded measurable h whose set of discontinuity points is a μ-null set.*

For a proof, see, for instance, Billingsley [24, Theorem 2.1]. The equivalence with the last condition can be found in Pollard [104, Section III.2].

1.7.1 Stability of Transference Plans and Cyclical Monotonicity

In this subsection, we state and sketch the proof of the fact that if $\mu_n \to \mu$ and $\nu_n \to \nu$ weakly, then the optimal transference plans $\pi_n \in \Pi(\mu_n, \nu_n)$ converge to an optimal $\pi \in \Pi(\mu, \nu)$. The result, as stated in Villani [125, Theorem 5.20], is valid on complete separate metric spaces with general cost functions, and reads as follows.

Theorem 1.7.2 (Weak Convergence and Optimal Plans) *Let μ_n and ν_n converge weakly to μ and ν, respectively, in $P(\mathcal{X})$ and let $c : \mathcal{X}^2 \to \mathbb{R}_+$ be continuous. If $\pi_n \in \Pi(\mu_n, \nu_n)$ are optimal transference plans and*

$$\limsup_{n\to\infty} \int_{\mathcal{X}^2} c(x,y)\,\mathrm{d}\pi_n(x,y) < \infty.$$

then (π_n) is a tight sequence and each of its weak limits $\pi \in \Pi(\mu, \nu)$ is optimal.

One can even let c vary with n under some conditions.

Let $c(x,y) = \|x-y\|^2/2$. We prefer to keep the notation $c(\cdot,\cdot)$ in order to stress the generality of the arguments. A key idea in the proof is the replacement of optimality by another property called *cyclical monotonicity*, which behaves nicely with respect to weak convergence. To motivate this property, we recall the discrete case of Sect. 1.3 where $\mu = N^{-1}\sum_{i=1}^N \delta\{x_i\}$ and $\nu = N^{-1}\sum_{i=1}^N \delta\{y_i\}$. There exists an optimal transference plan π induced from a permutation $\sigma_0 \in S_N$. Since the ordering of $\{x_i\}$ and $\{y_i\}$ is irrelevant in the representations of μ and ν, we may assume without loss of generality that σ_0 is the identity permutation. Then, by definition of optimality,

$$\sum_{i=1}^N c(x_i, y_i) \le \sum_{i=1}^N c(x_i, y_{\sigma(i)}), \qquad \sigma \in S_N. \tag{1.7}$$

If σ is the identity except for a subset $i_1, \dots, i_n$, $n \le N$, then in particular

$$\sum_{k=1}^n c(x_{i_k}, y_{i_k}) \le \sum_{k=1}^n c(x_{i_k}, y_{\sigma(i_k)}), \qquad \sigma \in S_n,$$

and if we choose $\sigma(i_k) = i_{k-1}$ with $i_0 = i_n$, this writes

$$\sum_{k=1}^n c(x_{i_k}, y_{i_k}) \le \sum_{k=1}^n c(x_{i_k}, y_{i_{k-1}}). \tag{1.8}$$

By decomposing a permutation $\sigma \in S_N$ to disjoint cycles, one can verify that (1.8) implies (1.7). This will be useful since, as it turns out, a variant of (1.8) holds for arbitrary measures μ and ν for which there is no relevant finite N as in (1.7).

Definition 1.7.3 (Cyclically Monotone Sets and Measures) *A set $\Gamma \subseteq \mathscr{X}^2$ is cyclically monotone if for any n and any $(x_1, y_1), \dots, (x_n, y_n) \in \Gamma$,*

$$\sum_{i=1}^{n} c(x_i, y_i) \leq \sum_{i=1}^{n} c(x_i, y_{i-1}), \qquad (y_0 = y_n). \tag{1.9}$$

A probability measure π on $\mathscr{X}^2$ is cyclically monotone if there exists a monotone Borel set Γ such that $\pi(\Gamma) = 1$.

The relevance of cyclical monotonicity becomes clear from the following observation. If μ and ν are discrete uniform measures on N points and σ is an optimal permutation for the Monge–Kantorovich problem, then the coupling $\pi = (1/N)\sum_{i=1}^{N} \delta\{(x_i, y_{\sigma(i)})\}$ is cyclically monotone. In fact, even if the optimal permutation is not unique, the set

$$\Gamma = \{(x_i, y_{\sigma(i)}) : i = 1, \dots, N, \sigma \in S_N \text{ optimal}\}$$

is cyclically monotone. Furthermore, $\pi \in \Pi(\mu, \nu)$ is optimal if and only if it is cyclically monotone, if and only if $\pi(\Gamma) = 1$. It is heuristically easy to see that cyclical monotonicity is a necessary condition for optimality:

Proposition 1.7.4 (Optimal Plans Are Cyclically Monotone) *Let $\mu, \nu \in P(\mathscr{X})$ and suppose that the cost function c is nonnegative and continuous. Assume that the optimal $\pi \in \Pi(\mu, \nu)$ has a finite total cost. Then* $\operatorname{supp}\pi$ *is cyclically monotone. In particular, π is cyclically monotone.*

The idea of the proof is that if for some $(x_1, y_1), \dots, (x_n, y_n)$ in the support of π,

$$\sum_{i=1}^{n} c(x_i, y_i) > \sum_{i=1}^{n} c(x_i, y_{i-1}),$$

then by continuity of c, the same inequality holds on some balls of positive measure. One can then replace π by a measure having (x_i, y_{i-1}) rather than (x_i, y_i) in its support, and this measure will incur a lower cost than π. A rigorous proof can be found in Gangbo and McCann [59, Theorem 2.3].

Thus, optimal transference plans π solve infinitely many discrete Monge–Kantorovich problems emanating from their support. More precisely, for any finite collection $(x_i, y_i) \in \operatorname{supp}\pi$, $i = 1, \dots, N$ and any permutation $\sigma \in S_N$, (1.7) is satisfied. Therefore, the identity permutation is optimal between the measures $(1/N)\sum\delta\{x_i\}$ and $(1/N)\sum\delta\{y_j\}$.

In the same spirit as Γ defined above for the discrete case, one can strengthen Proposition 1.7.4 and prove existence of a cyclically monotone set Γ that includes the support of *any* optimal transference plan π: take $\Gamma = \cup \operatorname{supp}(\pi)$ for π optimal.

The converse of Proposition 1.7.4 also holds.

Proposition 1.7.5 (Cyclically Monotone Plans Are Optimal) *Let $\mu, \nu \in P(\mathscr{X})$, $c : \mathscr{X}^2 \to \mathbb{R}_+$ continuous and $\pi \in \Pi(\mu, \nu)$ a cyclically monotone measure with $C(\pi)$ finite. Then π is optimal in $\Pi(\mu, \nu)$.*

Let us sketch the proof in the quadratic case $c(x,y) = \|x-y\|^2/2$ and see how convexity comes into play. Straightforward algebra shows that (1.9) is equivalent, in the quadratic case, to

$$\sum_{i=1}^{n} \langle y_i, x_{i+1} - x_i \rangle \leq 0, \qquad (x_{n+1} = x_1). \tag{1.10}$$

Fix $(x_0, y_0) \in \Gamma = \operatorname{supp}\pi$ and define $\phi : \mathscr{X} \to \mathbb{R} \cup \{\infty\}$ by

$$\phi(x) = \sup\{\langle y_0, x_1 - x_0 \rangle + \cdots + \langle y_{m-1}, x_m - x_{m-1} \rangle$$
$$+ \langle y_m, x - x_m \rangle : m \in \mathbb{N}, \quad (x_i, y_i) \in \Gamma\}.$$

This function is defined as a supremum of affine functions, and is therefore convex and lower semicontinuous. Cyclical monotonicity of Γ implies that $\phi(x_0) = 0$, so ϕ is not identically infinite (it would have been so if Γ were not cyclically monotone). Straightforward computations show that Γ is included in the subdifferential of ϕ: y is a subgradient of ϕ at x when $(x,y) \in \Gamma$. Optimality of π then follows by weak duality, since π assigns full measure to the set of (x,y) such that $\phi(x) + \phi^*(y) = \langle x, y \rangle$; see (1.5) and the discussion around it.

The argument for more general costs follows similar lines and is sketched at the end of this subsection.

Given these intermediary results, it is now instructive to prove Theorem 1.7.2.

Proof (Proof of Theorem 1.7.2). Since $\mu_n \to \mu$ weakly, it is a tight sequence, and similarly for ν_n. Consequently, the entire set of plans $\cup_n \Pi(\mu_n, \nu_n)$ is tight too (see the discussion before deriving (1.3)). Therefore, up to a subsequence, (π_n) has a weak limit π. We need to show that π is cyclically monotone and that $C(\pi)$ is finite. The latter is easy, since $c_M(x,y) = \min(M, c(x,y))$ is continuous and bounded:

$$C(\pi) = \lim_{M\to\infty} \int_{\mathscr{X}^2} c_M \, d\pi = \lim_{M\to\infty} \lim_{n\to\infty} \int_{\mathscr{X}^2} c_M \, d\pi_n \leq \liminf_{n\to\infty} \int_{\mathscr{X}^2} c \, d\pi_n < \infty.$$

To show that π is cyclically monotone, fix $(x_1, y_1), \ldots, (x_N, y_N) \in \operatorname{supp}\pi$. We show that there exist $(x_k^n, y_k^n) \in \operatorname{supp}\pi_n$ that converge to (x_k, y_k). Once this is established, we conclude from the cyclical monotonicity of $\operatorname{supp}\pi_n$ and the continuity of c that

$$\sum_{k=1}^{N} c(x_k, y_k) = \lim_{n\to\infty} \sum_{k=1}^{N} c(x_k^n, y_k^n) \leq \lim_{n\to\infty} \sum_{k=1}^{N} c(x_k^n, y_{k-1}^n) = \sum_{k=1}^{N} c(x_k, y_{k-1}).$$

The existence proof for the sequence is standard. For $\varepsilon > 0$, let $B = B_\varepsilon(x_k, y_k)$ be an open ball around (x_k, y_k). Then $\pi(B) > 0$ and by the portmanteau Lemma 1.7.1, $\pi_n(B) > 0$ for sufficiently large n. It follows that there exist $(x_k^n, y_k^n) \in B \cap \operatorname{supp}\pi_n$. Let $\varepsilon = 1/m$, say, then for all $n \geq N_m$ we can find $(x_k^n, y_k^n) \in \operatorname{supp}\mu_n$ of distance $2/m$ from (x_k, y_k). We can choose $N_{m+1} > N_m$ without loss of generality in order to complete the proof.

A few remarks are in order. Firstly, quadratic cyclically monotone sets (with respect to $\|x-y\|^2/2$) are included in the subdifferential of convex functions. The converse is also true, as can be easily deduced from summing up the subgradient inequalities

$$\phi(x_{i+1}) \geq \phi(x_i) + \langle y_i, x_{i+1} - x_i \rangle, \qquad i = 1, \ldots, N,$$

where y_i is a subgradient of ϕ at x_i. For future reference, we state this characterisation as a theorem (which is valid in infinite dimensions too).

Theorem 1.7.6 (Rockafellar [112]) *A nonempty $\Gamma \subseteq \mathscr{X}^2$ is quadratic cyclically monotone if and only if it is included in the graph of the subdifferential of a lower semicontinuous convex function that is not identically infinite.*

Secondly, we have not used at all the Kantorovich duality, merely its weak form. The machinery of cyclical monotonicity can be used in order to prove the duality Theorem 1.4.2. This is indeed the strategy of Villani [125, Chapter 5], who explains its advantage with respect to Hahn–Banach-type duality proofs.

Lastly, the idea of the proof of Proposition 1.7.5 generalises to other costs in a natural way. Given a cyclically monotone (with respect to a cost function c) set Γ and a fixed pair $(x_0, y_0) \in \Gamma$, define (Rüschendorf [116])

$$\varphi(x) = \inf\{c(x_1, y_0) - c(x_0, y_0) + c(x_m, y_{m-1}) - c(x_{m-1}, y_{m-1}) + c(x, y_m) - c(x_m, y_m)\}.$$

Then under some conditions, (φ, ψ) is dual optimal for some ψ. As explained in Sect. 1.8, ψ can be chosen to be essentially φ^c (as defined in that section).

1.7.2 Stability of Transport Maps

We now extend the weak convergence of π_n to π of the previous subsection to convergence of optimal maps. Because of the applications we have in mind, we shall work exclusively in the Euclidean space $\mathscr{X} = \mathbb{R}^d$ with the quadratic cost function; our results can most likely be extended to more general situations.

In this setting, we know that optimal plans are supported on graphs of subdifferentials of convex functions. Suppose that π_n is induced by T_n and π is induced by T. Then in some sense, the weak convergence of π_n to π yields convergence of the graphs of T_n to the graph of T. Our goal is to strengthen this to uniform convergence of T_n to T. Roughly speaking, we show the following: there exists a set A with $\mu(A) = 1$ and such that T_n converge uniformly to T on every compact subset of A. For the reader's convenience, we give a user-friendly version here; a more general statement is given in Proposition 1.7.11 below.

Theorem 1.7.7 (Uniform Convergence of Optimal Maps) *Let μ_n, μ be absolutely continuous measures with finite second moments on an open convex set $U \subseteq \mathbb{R}^d$ such that $\mu_n \to \mu$ weakly, and let $\nu_n \to \nu$ weakly with $\nu_n, \nu \in P(\mathbb{R}^d)$ with finite second moments. If T_n and T are continuous on U and $C(T_n)$ is bounded uniformly in n, then*

$$\sup_{x\in\Omega}\|T_n(x)-T(x)\|\to 0, \qquad n\to\infty,$$

for any compact $\Omega\subseteq U$.

Since T_n and T are only defined up to Lebesgue null sets, it will be more convenient to work directly with the subgradients. That is, we view T_n and T as *set-valued* functions that to each $x\in\mathbb{R}^d$ assign a (possibly empty) subset of $\mathbb{R}^d$. In other words, T_n and T take values in the power set of $\mathbb{R}^d$, denoted by $2^{\mathbb{R}^d}$.

Let $\phi:\mathbb{R}^d\to\mathbb{R}\cup\{\infty\}$ be convex, $y_1\in\partial\phi(x_1)$ and $y_2\in\partial\phi(x_2)$. Putting $n=2$ in the definition of cyclical monotonicity (1.10) gives

$$\langle y_2-y_1,x_2-x_1\rangle\geq 0.$$

This property (which is weaker than cyclical monotonicity) is important enough to have its own name. Following the notation of Alberti and Ambrosio [6], we call a set-valued function (or multifunction) $u:\mathbb{R}^d\to 2^{\mathbb{R}^d}$ *monotone* if whenever $y_i\in u(x_i)$, $i=1,2$,

$$\langle y_2-y_1,x_2-x_1\rangle\geq 0.$$

If $d=1$, this simply means that u is a nondecreasing (set-valued) function. For example, one can define $u(x)=\{0\}$ for $x\in[0,1)$, $u(1)=[0,1]$ and $u(x)=\emptyset$ if $x\notin[0,1]$. Next, u is said to be *maximally monotone* if no points can be added to its graph while preserving monotonicity:

$$\{\langle y'-y,x'-x\rangle\geq 0 \quad \text{whenever } y\in u(x)\} \quad\Longrightarrow\quad y'\in u(x').$$

It will be convenient to identify u with its graph; we will often write $(x,y)\in u$ to mean $y\in u(x)$. Note that $u(x)$ can be empty, even when u is maximally monotone. The previous example for u is not maximally monotone, but it will be if we modify $u(0)$ to be $(-\infty,0]$ and $u(1)$ to be $[0,\infty)$.

Of course, if $\phi:\mathbb{R}^d\to\mathbb{R}\cup\{\infty\}$ is convex, then $u=\partial\phi$ is monotone. It follows from Theorem 1.7.6 that u is maximally cyclically monotone (no points can be added to its graph while preserving cyclical monotonicity). It can actually be shown that u is maximally monotone [6, Section 7]. In what follows, we will always work with subdifferentials of convex functions, so unless stated otherwise, u will always be assumed maximally monotone.

Maximally monotone functions enjoy the following very useful continuity property. It is proven in [6, Corollary 1.3] and will be used extensively below.

Proposition 1.7.8 (Continuity at Singletons) *Let* $x\in\mathbb{R}^d$ *such that* $u(x)=\{y\}$ *is a singleton. Then* u *is nonempty on some neighbourhood of* x *and it is continuous at* x*: if* $x_n\to x$ *and* $y_n\in u(x_n)$*, then* $y_n\to y$.

Notice that this result implies that if a convex function ϕ is differentiable on some open set $E\subseteq\mathbb{R}^d$, then it is continuously differentiable there (Rockafellar [113, Corollary 25.5.1]).

If $f : \mathbb{R}^d \to \mathbb{R} \cup \{\infty\}$ is any function, one can define its subgradient at x locally as

$$\partial f(x) = \{y : f(z) \geq f(x) + \langle y, z-x\rangle + o(\|z-x\|)\}$$
$$= \left\{y : \liminf_{z\to x} \frac{f(z) - f(x) + \langle y, z-x\rangle}{\|z-x\|} \geq 0\right\}.$$

(See the discussion after Theorem 1.8.2.) When f is convex, one can remove the $o(\|z-x\|)$ term and the inequality holds for all z, i.e., globally and not locally. Since monotonicity is related to convexity, it should not be surprising that monotonicity is in some sense a local property. Suppose that $u(x_0) = \{y_0\}$ is a singleton and that for some $y^* \in \mathbb{R}^d$,

$$\langle y - y^*, x - x_0\rangle \geq 0$$

for all $x \in \mathbb{R}^d$ and $y \in u(x)$. Then by maximality, y^* must equal y_0. By "local property", we mean that the conclusion $y^* = y_0$ holds if the above inequality holds for x in a small neighbourhood of x_0 (an open set that includes x_0). We will need a more general version of this result, replacing neighbourhoods by a weaker condition that can be related to Lebesgue points. The strengthening is somewhat technical; the reader can skip directly to Lemma 1.7.10 and assume that G is open without losing much intuition.

Let $B_r(x_0) = \{x : \|x - x_0\| < r\}$ for $r \geq 0$ and $x_0 \in \mathbb{R}^d$. The interior of a set $G \subseteq \mathbb{R}^d$ is denoted by $\mathrm{int}G$ and the closure by $\overline{G}$. If G is measurable, then $\mathrm{Leb}G$ denotes the Lebesgue measure of G. Finally, $\mathrm{conv}G$ denotes the convex hull of G.

A point x_0 is a *Lebesgue point* (or of *Lebesgue density*) of a measurable set $G \subseteq \mathbb{R}^d$ if for any $\varepsilon > 0$ there exists $t_\varepsilon > 0$ such that

$$\frac{\mathrm{Leb}(B_t(x_0) \cap G)}{\mathrm{Leb}(B_t(x_0))} > 1 - \varepsilon, \qquad 0 < t < t_\varepsilon.$$

An illuminating example is the set $\{y \leq \sqrt{|x|}\}$ in $\mathbb{R}^2$ (see Fig. 1.1). Since the "slope" of the square root is infinite, $x_0 = (0,0)$ is a Lebesgue point, but the fraction above is strictly smaller than one, for all $t > 0$.

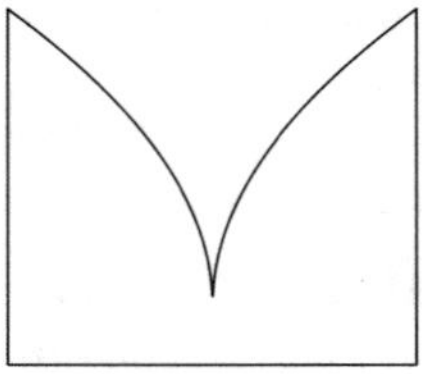

Fig. 1.1: The set $G = \{(x,y) : |x| \leq 1,\ -0.2 \leq y \leq \sqrt{|x|}\}$

We denote the set of points of Lebesgue density of G by G^{den}. Here are some facts about G^{den}: clearly, $\mathrm{int}G \subseteq G^{\mathrm{den}} \subseteq \overline{G}$. Stein and Shakarchi [121, Chapter 3, Corollary 1.5] show that $\mathrm{Leb}(G \setminus G^{\mathrm{den}}) = 0$ (and $\mathrm{Leb}(G^{\mathrm{den}} \setminus G) = 0$, so G^{den} is very close to G). By the Hahn–Banach theorem, $G^{\text{den}} \subseteq \text{int}(\text{conv}(G))$: indeed, if x is not in $\text{int}(\text{conv}G)$, then there is a separating hyperplane between x and $\text{conv}G \supseteq G$, so the fraction above is at most $1/2$ for all $t > 0$.

The "denseness" of Lebesgue points is materialised in the following result. It is given as exercise in [121] when $d = 1$, and the proof can be found on page 27 in the supplement.

Lemma 1.7.9 (Density Points and Distance) *Let x_0 be a point of Lebesgue density of a measurable set $G \subseteq \mathbb{R}^d$. Then*

$$\delta(z) = \delta_G(z) = \inf_{x \in G} \|z - x\| = o(\|z - x_0\|), \qquad \textit{as } z \to x_0.$$

Of course, this result holds for any $x_0 \in \overline{G}$ if the little o is replaced by big O, since δ is Lipschitz. When $x_0 \in \text{int}G$, this is trivial because δ vanishes on $\text{int}G$.

The important part here is the following corollary: for almost all $x \in G$, $\delta(z) = o(\|z - x\|)$ as $z \to x$. This can be seen in other ways: since δ is Lipschitz, it is differentiable almost everywhere. If $x \in \overline{G}$ and δ is differentiable at x, then $\nabla\delta(x)$ must be 0 (because δ is minimised there), and then $\delta(z) = o(\|z - x\|)$. We just showed that δ is differentiable with vanishing derivative at all Lebesgue points of x. The converse is not true: $G = \{\pm 1/n\}_{n=1}^{\infty}$ has no Lebesgue points, but $\delta(y) \le 4y^2$ as $y \to 0$.

The locality of monotone functions can now be stated as follows. It is proven on page 27 of the supplement.

Lemma 1.7.10 (Local Monotonicity) *Let $x_0 \in \mathbb{R}^d$ such that $u(x_0) = \{y_0\}$ and x_0 is a Lebesgue point of a set G satisfying*

$$\langle y - y^*, x - x_0 \rangle \ge 0 \qquad \forall x \in G\ \forall y \in u(x).$$

Then $y^* = y_0$. In particular, the result is true if the inequality holds on $G = O \setminus \mathscr{N}$ with $\emptyset \ne O$ open and $\mathscr{N}$ Lebesgue negligible.

These continuity properties cannot be of much use unless $u(x)$ is a singleton for reasonably many values of x. Fortunately, this is indeed the case: the set of points x such that $u(x)$ contains more than one element has Lebesgue measure 0 (see Alberti and Ambrosio [6, Remark 2.3] for a stronger result). Another issue is that u may be empty, and convexity comes into play here again. Let $\text{dom}u = \{x : u(x) \ne \emptyset\}$. Then there exists a convex closed set K such that

$$\text{int}K \subseteq \text{dom}u \subseteq K.$$

[6, Corollary 1.3(2)]. Although $\text{dom}u$ itself may fail to be convex, it is almost convex in the above sense. By convexity, $K \setminus \text{int}K$ has Lebesgue measure 0 (see the discussion after Theorem 1.6.1) and so the set of points in K where u is not a singleton,

$$\{x \in K : u(x) = \emptyset\} \cup \{x \in K : u(x) \text{ contains more than one point}\},$$

has Lebesgue measure 0, and $u(x)$ is empty for all $x \notin K$. (It is in fact not difficult to show that if $x \in \partial K$, then $u(x)$ cannot be a singleton, by the Hahn–Banach theorem.)

With this background on monotone functions at our disposal, we are now ready to state the stability result for the optimal maps. We assume the following.

Assumptions 1 *Let $\mu_n, \mu, \nu_n, \nu \in P(\mathbb{R}^d)$ with optimal couplings (with respect to quadratic cost) $\pi_n \in \Pi(\mu_n, \nu_n)$, $\pi \in \Pi(\mu, \nu)$ and convex potentials ϕ_n and ϕ, respectively, such that*

- (convergence) *$\mu_n \to \mu$ and $\nu_n \to \nu$ weakly;*
- (finiteness) *the optimal couplings $\pi_n \in \Pi(\mu_n, \nu_n)$ satisfy*

$$\limsup_{n\to\infty} \int_{\mathcal{X}^2} \frac{1}{2}\|x-y\|^2 \, \mathrm{d}\pi_n(x,y) < \infty;$$

- (unique limit) *the optimal $\pi \in \Pi(\mu, \nu)$ is unique.*

We further denote the subgradients $\partial\phi_n$ and $\partial\phi$ by u_n and u, respectively.

These assumptions imply that π has a finite total cost. This can be shown by the liminf argument in the proof of Theorem 1.7.2 but also from the uniqueness of π. As a corollary of the uniqueness of π, it follows that $\pi_n \to \pi$ weakly; notice that this holds even if π_n is not unique for any n. We will now translate this weak convergence to convergence of the maximal monotone maps u_n to u, in the following form.

Proposition 1.7.11 (Uniform Convergence of Optimal Maps) *Let Assumptions 1 hold true and denote $E = \operatorname{supp}\mu$ and E^{den} the set of its Lebesgue points. Let Ω be a compact subset of E^{den} on which u is univalued (i.e., $u(x)$ is a singleton for all $x \in \Omega$). Then u_n converges to u uniformly on Ω: $u_n(x)$ is nonempty for all $x \in \Omega$ and all $n > N_\Omega$, and*

$$\sup_{x\in\Omega} \sup_{y\in u_n(x)} \|y - u(x)\| \to 0, \qquad n \to \infty.$$

In particular, if u is univalued throughout $\operatorname{int}(E)$ *(so that $\phi \in C^1$ there), then uniform convergence holds for any compact $\Omega \subset \operatorname{int}(E)$.*

The proof of Proposition 1.7.11, given on page 28 of the supplement, follows two separate steps:

- if a sequence in the graph of u_n converges, then the limit is in the graph of u;
- sequences in the graph of u_n are bounded if the domain is bounded.

Corollary 1.7.12 (Pointwise Convergence μ-Almost Surely) *If in addition μ is absolutely continuous, then $u_n(x) \to u(x)$ μ-almost surely.*

Proof. We first claim that $E \subseteq \overline{\operatorname{dom} u}$. Indeed, for any $x \in E$ and any $\varepsilon > 0$, the ball $B = B_\varepsilon(x)$ has positive measure. Consequently, u cannot be empty on the entire ball, because otherwise $\mu(B) = \pi(B \times \mathbb{R}^d)$ would be 0. Since $\operatorname{dom} u$ is almost convex (see the discussion before Assumptions 1), this implies that actually $\operatorname{int}(\operatorname{conv} E) \subseteq \operatorname{dom} u$.

The rest is now easy: the set of points $x \in E$ for which $\Omega = \{x\}$ fails to satisfy the conditions of Proposition 1.7.11 is included in

$$(E \setminus E^{\mathrm{den}}) \cup \{x \in \mathrm{int}(\mathrm{conv}(E)) : u(x) \text{ contains more than one point}\},$$

which is μ-negligible because μ is absolutely continuous and both sets have Lebesgue measure 0.

1.8 Complementary Slackness and More General Cost Functions

It is well-known (Luenberger and Ye [89, Section 4.4]) that the solutions to the primal and dual problems are related to each other via *complementary slackness*. In other words, solution of one problem provides a lot of information about the solution of the other problem. Here, we show that this idea remains true for the Kantorovich primal and dual problems, extending the discussion in Sect. 1.6.1 to more general cost functions.

Let $\mathscr{X}$ and $\mathscr{Y}$ be complete separable metric spaces, $\mu \in P(\mathscr{X})$, $\nu \in P(\mathscr{Y})$, and $c : \mathscr{X} \times \mathscr{Y} \to \mathbb{R}_+$ be a measurable cost function.

If one finds functions $(\varphi, \psi) \in \Phi_c$ and a transference plan $\pi \in \Pi(\mu, \nu)$ having the same objective values, then by weak duality (φ, ψ) is optimal in Φ_c and π is optimal in $\Pi(\mu, \nu)$. Having the same objective values is equivalent to

$$\int_{\mathscr{X} \times \mathscr{Y}} [c(x,y) - \varphi(x) - \psi(y)] \, \mathrm{d}\pi(x,y) = 0$$

which is in turn equivalent to

$$\varphi(x) + \psi(y) = c(x,y), \qquad \pi\text{-almost surely.}$$

It has already been established that there exists an optimal transference plan π^*. Assuming that $C(\pi^*) < \infty$ (otherwise all transference plans are optimal), a pair $(\varphi, \psi) \in \Phi_c$ is optimal if and only if

$$\varphi(x) + \psi(y) = c(x,y), \qquad \pi^*\text{-almost surely.}$$

Conversely, if (φ_0, ψ_0) is an optimal pair, then π is optimal if and only if it is concentrated on the set

$$\{(x,y) : \varphi_0(x) + \psi_0(y) = c(x,y)\}.$$

In particular, if for a given x there exists a unique y such that $\varphi_0(x) + \psi_0(y) = c(x,y)$, then the mass at x must be sent entirely to y and not be split; if this is the case for μ-almost all x, then this relation defines y as a function of x and the resulting optimal π is in fact induced from a transport map. This idea provides a criterion for solvability of the Monge problem (Villani [125, Theorem 5.30]).

1.8.1 Unconstrained Dual Kantorovich Problem

It turns out that the dual Kantorovich problem can be recast as an unconstrained optimisation problem of only one function φ. The new formulation is not only conceptually simpler than the original one, but also sheds light on the properties of the optimal dual variables. Since the dual objective function to be maximised,

$$\int_{\mathcal{X}} \varphi \, d\mu + \int_{\mathcal{Y}} \psi \, d\nu,$$

is increasing in φ and ψ, one should seek functions that take values as large as possible subject to the constraint $\varphi(x) + \psi(y) \leq c(x,y)$. Suppose that an oracle tells us that some $\varphi \in L_1(\mu)$ is a good candidate. Then the largest possible ψ satisfying $(\varphi, \psi) \in \Phi_c$ is defined as

$$\psi(y) = \inf_{x \in \mathcal{X}} [c(x,y) - \varphi(x)] := \varphi^c(y).$$

A function taking this form is called *c-concave* [124, Chapter 2]; we say that ψ is the *c-transform* of φ. It is not necessarily true that φ^c is integrable or even measurable, but if we neglect this difficulty, then it is obvious that

$$\sup_{\psi \in L_1(\nu):(\varphi,\psi)\in\Phi_c} \left[\int_{\mathcal{X}} \varphi \, d\mu + \int_{\mathcal{Y}} \psi \, d\nu \right] = \int_{\mathcal{X}} \varphi \, d\mu + \int_{\mathcal{Y}} \varphi^c \, d\nu.$$

The dual problem can thus be formulated as the unconstrained problem

$$\sup_{\varphi \in L_1(\mu)} \left[\int_{\mathcal{X}} \varphi \, d\mu + \int_{\mathcal{Y}} \varphi^c \, d\nu \right].$$

One can apply this c-transform again and replace φ by

$$\varphi^{cc}(x) = (\varphi^c)^c(x) = \inf_{y \in \mathcal{Y}} [c(x,y) - \varphi^c(y)] \geq \varphi(x),$$

so that φ^{cc} has a better objective value yet still $(\varphi^{cc}, \varphi^c) \in \Phi_c$ (modulo measurability issues). An elementary calculation shows that in general $\varphi^{ccc} = \varphi^c$. Thus, for any function φ_1, the pair of functions $(\varphi, \psi) = (\varphi_1^{cc}, \varphi_1^c)$ has a better objective value than (φ_1, ψ_1), and satisfies $(\varphi, \psi) \in \Phi_c$. Moreover, $\varphi^c = \psi$ and $\psi^c = \varphi$; in words, φ and ψ are *c-conjugate*. An optimal dual pair (φ, ψ) can be expected to be c-conjugate; this is indeed true almost surely:

Proposition 1.8.1 (Existence of an Optimal Pair) *Let μ and ν be probability measures on $\mathcal{X}$ and $\mathcal{Y}$ such that the independent coupling with respect to the nonnegative and lower semicontinuous cost function is finite: $\int_{\mathcal{X}\times\mathcal{Y}} c(x,y) \, d\mu(x) d\nu(y) < \infty$. Then there exists an optimal pair (φ, ψ) for the dual Kantorovich problem. Furthermore, the pair can be chosen in a way that μ-almost surely, $\varphi = \psi^c$ and ν-almost surely, $\psi = \varphi^c$.*

Proposition 1.8.1 is established (under weaker conditions) by Ambrosio and Pratelli [11, Theorem 3.2]. It is clear from the discussion above that once existence of an optimal pair (φ_1, ψ_1) is established, the pair $(\varphi, \psi) = (\varphi_1^{cc}, \varphi_1^c)$ should be optimal. Combining Proposition 1.8.1 with the preceding subsection, we see that if φ is optimal (for the unconstrained dual problem), then any optimal transference plan π^* must be concentrated on the set

$$\{(x,y) : \varphi(x) + \varphi^c(y) = c(x,y)\}.$$

If for μ-almost every x this equation defines y uniquely as a (measurable) function of x, then π^* is induced by a transport map. Indeed, we have seen how this is the case, in the quadratic case $c(x,y) = \|x-y\|^2/2$, when μ is absolutely continuous. An extension to $p > 1$ (instead of $p = 2$) is sketched in Sect. 1.8.3.

We remark that at the level of generality of Proposition 1.8.1, the function φ^c may fail to be Borel measurable; Ambrosio and Pratelli show that this pair can be modified up to null sets in order to be Borel measurable. If c is continuous, however, then φ^c is an infimum of a collection of continuous functions (in y). Hence $-\varphi^c$ is lower semicontinuous, which yields that φ^c is measurable. When c is *uniformly* continuous, measurability of φ^c is established in a more lucid way, as exemplified in the next subsection.

1.8.2 The Kantorovich–Rubinstein Theorem

Whether $\varphi^c(y)$ is tractable to evaluate depends on the structure of c. We have seen an example where c was the quadratic Euclidean distance. Here, we shall consider another useful case, where c is a metric. Assume that $\mathscr{X} = \mathscr{Y}$, denote their metric by d, and let $c(x,y) = d(x,y)$. If $\varphi = \psi^c$ is c-concave, then it is 1-Lipschitz. Indeed, by definition and the triangle inequality

$$\varphi(z) = \inf_{y\in\mathscr{Y}} [d(z,y) - \psi(y)] \le \inf_{y\in\mathscr{Y}} [d(x,y) + d(x,z) - \psi(y)] = \varphi(x) + d(x,z).$$

Interchanging x and z yields $|\varphi(x) - \varphi(z)| \le d(x,z)$.[3]

Next, we claim that if φ is Lipschitz, then $\varphi^c(y) = -\varphi(y)$. Indeed, choosing $x = y$ in the infimum shows that $\varphi^c(y) \le d(y,y) - \varphi(y) = -\varphi(y)$. But the Lipschitz condition on φ implies that for all x, $d(x,y) - \varphi(x) \ge -\varphi(y)$. In view of that, we can take in the dual problem φ to be Lipschitz and $\psi = -\varphi$, and the duality formula (Theorem 1.4.2) takes the form

$$\inf_{\pi\in\Pi(\mu,\nu)} \int_{\mathscr{X}^2} d(x,y)\,\mathrm{d}\pi(x,y) = \sup_{\|\varphi\|_{\mathrm{Lip}}\le 1} \left| \int_{\mathscr{X}} \varphi\,\mathrm{d}\mu - \int_{\mathscr{X}} \varphi\,\mathrm{d}\nu \right|,$$

$$\|\varphi\|_{\mathrm{Lip}} = \sup_{x\ne y} \frac{|\varphi(x) - \varphi(y)|}{d(x,y)}. \tag{1.11}$$

---

[3] In general, ψ^c inherits the modulus of continuity of c, see Santambrogio [119, page 11].

This is known as the *Kantorovich–Rubinstein theorem* [124, Theorem 1.14]. (We have been a bit sloppy because φ may not be integrable. But if for some $x_0 \in \mathscr{X}$, $x \mapsto d(x,x_0)$ is in $L_1(\mu)$, then any Lipschitz function is μ-integrable. Otherwise one needs to restrict the supremum to, e.g., bounded Lipschitz φ.)

1.8.3 Strictly Convex Cost Functions on Euclidean Spaces

We now return to the Euclidean case $\mathscr{X} = \mathscr{Y} = \mathbb{R}^d$ and explore the structure of c-transforms. When c is different than $\|x-y\|^2/2$, we can no longer "open up the square" and relate the Monge–Kantorovich problem to convexity. However, we can still apply the idea that $\varphi(x) + \varphi^c(y) = c(x,y)$ if and only if the infimum is attained at x. Indeed, recall that

$$\varphi^c(y) = \inf_{x\in\mathscr{X}} [c(x,y) - \varphi(x)],$$

so that $\varphi(x) + \varphi^c(y) = c(x,y)$ if and only if

$$\varphi(z) - \varphi(x) \le c(z,y) - c(x,y), \qquad z \in \mathscr{X}.$$

Notice the similarity to the subgradient inequality for convex functions, with the sign being reversed. In analogy, we call the collection of y's satisfying the above in equality the *c-superdifferential* of φ at x, and we denote it by $\partial^c\varphi(x)$. Of course, if $c(x,y) = \|x-y\|^2/2$, then $y \in \partial^c(x)$ if and only if y is a subgradient of $(\|\cdot\|^2/2 - \varphi)$ at x.

The following result generalises Theorem 1.6.2 to other powers $p > 1$ of the Euclidean norm. These cost functions define the Wasserstein distances of the next chapter.

Theorem 1.8.2 (Strictly Convex Costs on $\mathbb{R}^d$) *Let $c(x,y) = h(x-y)$ with $h(v) = \|v\|^p/p$ for some $p > 1$ and let μ and ν be probability measures on $\mathbb{R}^d$ with finite p-th moments such that μ is absolutely continuous with respect to Lebesgue measure. Then the solution to the Kantorovich problem with cost function c is unique and induced from a transport map T. Furthermore, there exists an optimal pair (φ,φ^c) of the dual problem, with φ c-concave. The solutions are related by*

$$T(x) = x - \nabla\varphi(x)\|\nabla\varphi(x)\|^{1/(p-1)-1} \qquad (\mu\text{-almost surely}).$$

Proof (Assuming ν has Compact Support). The existence of the optimal pair (φ,φ^c) with the desired properties follows from Proposition 1.8.1 (they are Borel measurable because c is continuous). We shall now show that φ has a unique c-supergradient μ-almost surely.

Step 1: φ is c-superdifferentiable. Let π^* be an optimal coupling. By duality arguments, π is concentrated on the set of (x,y) such that $y \in \partial^c\varphi(x)$. Consequently, for μ-almost any x, the c-superdifferential of φ at x is nonempty.

Step 2: φ is differentiable. Here, we impose the additional condition that ν is compactly supported. Then φ can be taken as a c-transform on the compact support of ν. Since h is locally Lipschitz (it is C^1 because $p > 1$) this implies that φ is locally Lipschitz. Hence, it is differentiable Lebesgue almost surely, and consequently μ-almost surely.

Step 3: Conclusion. For μ-almost every x there exists $y \in \partial^c\varphi(x)$ and a gradient $u = \nabla\varphi(x)$. In particular, u is a subgradient of φ:

$$\varphi(z) - \varphi(x) \geq \langle u, z - x\rangle + o(\|z - x\|).$$

Here and more generally, $o(\|z-x\|)$ denotes a function $r(z)$ (defined in a neighbourhood of x) such that $r(z)/\|z-x\| \to 0$ as $z \to x$. (If φ were convex, then we could take $r \equiv 0$, so the definition for convex functions is equivalent, and then the inequality holds globally and not only locally.) But $y \in \partial^c\varphi(x)$ means that as $z \to x$,

$$h(z-y) - h(x-y) = c(z,y) - c(x,y) \geq \varphi(z) - \varphi(x) \geq \langle u, z-x\rangle + o(\|z-x\|).$$

In other words, u is a subgradient of h at $x - y$. But h is differentiable with gradient $\nabla h(v) = v\|v\|^{p-2}$ (zero if $v = 0$). We obtain $\nabla\varphi(x) = u = \nabla h(x-y)$ and since the gradient of h is invertible, we conclude

$$y = T(x) := x - (\nabla h)^{-1}[\nabla\varphi(x)],$$

which defines y as a (measurable) function of x.[4] Hence, the optimal transference plan π is unique and induced from the transport map T.

The general result, without assuming compact support for ν, can be found in Gangbo and McCann [59]. It holds for a larger class of functions h, those that are strictly convex on $\mathbb{R}^d$ (this yields that ∇h is invertible), have superlinear growth ($h(v)/\|v\| \to \infty$ as $v \to \infty$) and satisfying a technical geometric condition (which $\|v\|^p/p$ does when $p > 1$). Furthermore, if h is sufficiently smooth, namely $h \in C^{1,1}$ locally (it is if $p \geq 2$, but not if $p \in (1,2)$), then μ does not need to be absolutely continuous; it suffices that it not give positive measure to any set of Hausdorff dimension smaller or equal than $d-1$. When $d = 1$ this means that Theorem 1.8.2 is still valid as long as μ has no atoms ($\mu(\{x\}) = 0$ for all $x \in \mathbb{R}$), which is a weaker condition than μ being absolutely continuous.

It is also noteworthy that for strictly concave cost functions (e.g., $p \in (0,1)$), the situation is similar *when the supports of μ and ν are disjoint*. The reason is that h may fail to be differentiable at 0, but it only needs to be differentiated at $x - y$ with $x \in \operatorname{supp}\mu$ and $y \in \operatorname{supp}\nu$. If the supports are not disjoint, then one needs to leave all common mass in place until the supports become disjoint (Villani [124, Chapter 2]) and then the result of [59] applies. As a simple example, let μ be uniform on $[0,1]$ and ν be uniform on $[0,2]$. After leaving common mass in place, we are left with uniforms on $[0,1]$ and $[1,2]$ (with total mass $1/2$) with essentially disjoint supports,

[4] Gradients of Borel functions are measurable, as the limit can be taken on a countable set. The inverse $(\nabla h)^{-1}$ equals the gradient of the Legendre transform h^* and is therefore Borel measurable.

for which the optimal transport map is the *decreasing* map $T(x) = 2 - x$. Thus, the unique optimal π is not induced from a map, but rather from an equal weight mixture of T and the identity. Informally, each point x in the support of μ needs to be split; half stays and x and the other half transported to $2 - x$. The optimal coupling from ν to μ is unique and induced from the map $S(x) = x$ if $x \leq 1$ and $2 - x$ if $x \geq 1$, which is neither increasing nor decreasing.

1.9 Bibliographical Notes

Many authors, including Villani [124, Theorem 1.3]; [125, Theorem 5.10], give the duality Theorem 1.4.2 for lower semicontinuous cost functions. The version given here is a simplification of Beiglböck and Schachermayer [17, Theorem 1]. The duality holds for functions that take values in $[-\infty, \infty]$ provided that they are finite on a sufficiently large subset of $\mathscr{X} \times \mathscr{Y}$, but there are simple counterexamples if c is infinite too often [17, Example 4.1]. For results outside the Polish space setup, see Kellerer [80] and Rachev and Rüschendorf [107, Chapter 4].

Theorem 1.5.1 for the one-dimensional case is taken from [124], where it is proven using the general duality theorem. For direct proofs and the history of this result, one may consult Rachev [106] or Rachev and Rüschendorf [107, Section 3.1]. The concave case is carefully treated by McCann [94].

The results in the Gaussian case were obtained independently by Olkin and Pukelsheim [98] and Givens and Shortt [65]. The proof given here is from Bhatia [20, Exercise 1.2.13]. An extension to separable Hilbert spaces can be found in Gelbrich [62] or Cuesta-Albertos et al. [39].

The regularity theory of Sect. 1.6.4 is very delicate. Caffarelli [32] showed the first part of Theorem 1.6.7; the proof can also be found in Figalli's book [52, Theorem 4.23]. Villani [124, Theorem 4.14] states the result without proof and refers to Alesker et al. [7] for a sketch of the second part of Theorem 1.6.7. Other regularity results exist, Villani [125, Chapter 12]; Santambrogio [119, Section 1.7.6]; Figalli [52].

Cuesta-Albertos et al. [40, Theorem 3.2] prove Theorem 1.7.2 for the quadratic case; the form given here is from Schachermayer and Teichmann [120, Theorem 3].

The definition of cyclical monotonicity depends on the cost function. It is typically referred to as c-cyclical monotonicity, with "cyclical monotonicity" reserved to the special case of quadratic cost. Since we focus on the quadratic case and for readability, we slightly deviate from the standard jargon. That cyclical monotonicity implies optimality (Proposition 1.7.5) was shown independently by Pratelli [105] (finite lower semicontinuous cost) and Schachermayer and Teichmann [120] (possibly infinite continuous cost). A joint generalisation is given by Beiglböck et al. [18].

Section 1.7.2 is taken from Zemel and Panaretos [134, Section 7.5]; a slightly weaker version was shown independently by Chernozhukov et al. [35]. Heinich and Lootgieter [68] establish almost sure pointwise convergence. If $\mu_n = \mu$, then the optimal maps converge in μ-measure [125, Corollary 5.23] in a very general setup, but there are simple examples where this fails if $\mu_n \neq \mu$ [125, Remark 5.25]. In the quadratic case, further stability results of a weaker flavour (focussing on the convex potential ϕ, rather than its derivative, which is the optimal map) can be found in del Barrio and Loubes [42].

The idea of using the c-transform (Sect. 1.8) is from Rüschendorf [116].

Chapter 2
The Wasserstein Space

The Kantorovich problem described in the previous chapter gives rise to a metric structure, the *Wasserstein distance*, in the space of probability measures $P(\mathscr{X})$ on a space $\mathscr{X}$. The resulting metric space, a subspace of $P(\mathscr{X})$, is commonly known as the *Wasserstein space* $\mathscr{W}$ (although, as Villani [125, pages 118–119] puts it, this terminology is "very questionable"; see also Bobkov and Ledoux [25, page 4]). In Chap. 4, we shall see that this metric is in a sense canonical when dealing with warpings, that is, deformations of the space $\mathscr{X}$ (for example, in Theorem 4.2.4). In this chapter, we give the fundamental properties of the Wasserstein space. After some basic definitions, we describe the topological properties of that space in Sect. 2.2. It is then explained in Sect. 2.3 how $\mathscr{W}$ can be endowed with a sort of infinite-dimensional Riemannian structure. Measurability issues are dealt with in the somewhat technical Sect. 2.4.

2.1 Definition, Notation, and Basic Properties

Let $\mathscr{X}$ be a separable Banach space. The *p-Wasserstein space* on $\mathscr{X}$ is defined as

$$\mathscr{W}_p(\mathscr{X}) = \left\{ \mu \in P(\mathscr{X}) : \int_{\mathscr{X}} \|x\|^p \, d\mu(x) < \infty \right\}, \qquad p \geq 1.$$

We will sometimes abbreviate and write simply $\mathscr{W}_p$ instead of $\mathscr{W}_p(\mathscr{X})$.

Recall that if $\mu, \nu \in P(\mathscr{X})$, then $\Pi(\mu, \nu)$ is defined to be the set of measures $\pi \in P(\mathscr{X}^2)$ having μ and ν as marginals in the sense of (1.2). The *p-Wasserstein distance* between μ and ν is defined as the minimal total transportation cost between

Electronic Supplementary Material The online version of this chapter (https://doi.org/10.1007/978-3-030-38438-8_2) contains supplementary material.

V. M. Panaretos, Y. Zemel, *An Invitation to Statistics in Wasserstein Space*, SpringerBriefs in Probability and Mathematical Statistics,
https://doi.org/10.1007/978-3-030-38438-8_2

μ and ν in the Kantorovich problem with respect to the cost function $c_p(x,y) = \|x-y\|^p$:

$$W_p(\mu,\nu) = \left(\inf_{\pi\in\Pi(\mu,\nu)} C_p(\pi)\right)^{1/p} = \left(\inf_{\pi\in\Pi(\mu,\nu)} \int_{\mathscr{X}\times\mathscr{X}} \|x_1-x_2\|^p \,\mathrm{d}\pi(x_1,x_2)\right)^{1/p}.$$

The Wasserstein distance between μ and ν is finite when both measures are in $\mathscr{W}_p(\mathscr{X})$, because

$$\|x_1-x_2\|^p \leq 2^p\|x_1\|^p + 2^p\|x_2\|^p.$$

Thus, W_p is finite on $[\mathscr{W}_p(\mathscr{X})]^2 = \mathscr{W}_p(\mathscr{X}) \times \mathscr{W}_p(\mathscr{X})$; it is nonnegative and symmetric and it is easy to see that $W_p(\mu,\nu) = 0$ if and only if $\mu = \nu$. A proof that W_p is a metric (satisfies the triangle inequality) on $\mathscr{W}_p$ can be found in Villani [124, Chapter 7].

The aforementioned setting is by no means the most general one can consider. Firstly, one can define W_p and $\mathscr{W}_p$ for $0 < p < 1$ by removing the power $1/p$ from the infimum and the limit case $p = 0$ yields the total variation distance. Another limit case can be defined as $W_\infty(\mu,\nu) = \lim_{p\to\infty} W_p(\mu,\nu)$. Moreover, W_p and $\mathscr{W}_p$ can be defined whenever $\mathscr{X}$ is a complete and separable metric space (or even only separable; see Clément and Desch [36]): one fixes some x_0 in $\mathscr{X}$ and replaces $\|x\|$ by $d(x,x_0)$. Although the topological properties below still hold at that level of generality (except when $p = 0$ or $p = \infty$), for the sake of simplifying the notation we restrict the discussion to Banach spaces. It will always be assumed without explicit mention that $1 \leq p < \infty$.

The space $\mathscr{W}_p(\mathscr{X})$ is defined as the collection of measures μ such that $W_p(\mu,\delta_0) < \infty$ with δ_x being a Dirac measure at x. Of course, $W_p(\mu,\nu)$ can be finite even if $\mu,\nu \notin \mathscr{W}_p(\mathscr{X})$. But if $\mu \in \mathscr{W}_p(\mathscr{X})$ and $\nu \notin \mathscr{W}_p(\mathscr{X})$, then $W_p(\mu,\nu)$ is always infinite. This can be seen from the triangle inequality

$$\infty = W_p(\nu,\delta_0) \leq W_p(\mu,\delta_0) + W_p(\mu,\nu).$$

In the sequel, we shall almost exclusively deal with measures in $\mathscr{W}_p(\mathscr{X})$.

The Wasserstein spaces are ordered in the sense that if $q \geq p$, then $\mathscr{W}_q(\mathscr{X}) \subseteq \mathscr{W}_p(\mathscr{X})$. This property extends to the distances in the form:

$$q \geq p \geq 1 \quad \Longrightarrow \quad W_q(\mu,\nu) \geq W_p(\mu,\nu). \tag{2.1}$$

To see this, let $\pi \in \Pi(\mu,\nu)$ be optimal with respect to q. Jensen's inequality for the convex function $z \mapsto z^{q/p}$ gives

$$W_q^q(\mu,\nu) = \int_{\mathscr{X}^2} \|x-y\|^q \,\mathrm{d}\pi(x,y) \geq \left(\int_{\mathscr{X}^2} \|x-y\|^p \,\mathrm{d}\pi(x,y)\right)^{q/p} \geq W_p^q(\mu,\nu).$$

The converse of (2.1) fails to hold in general, since it is possible that W_p be finite while W_q is infinite. A converse can be established, however, if μ and ν are bounded:

$$q\geq p\geq 1, \quad \mu(K)=\nu(K)=1 \quad \Longrightarrow \quad W_q(\mu,\nu)\leq W_p^{p/q}(\mu,\nu)\left(\sup_{x,y\in K}\|x-y\|\right)^{1-p/q}. \tag{2.2}$$

Indeed, if we denote the supremum by d_K and let π be now optimal with respect to p, then $\pi(K\times K)=1$ and

$$W_q^q(\mu,\nu)\leq \int_{K^2}\|x-y\|^q\,\mathrm{d}\pi(x,y)\leq d_K^{q-p}\int_{K^2}\|x-y\|^p\,\mathrm{d}\pi(x,y)=d_K^{q-p}W_p^p(\mu,\nu).$$

Another useful property of the Wasserstein distance is the upper bound

$$\mathscr{W}_p(\mathbf{t}\#\mu,\mathbf{s}\#\mu)\leq\left(\int_{\mathscr{X}}\|\mathbf{t}(x)-\mathbf{s}(x)\|^p\,\mathrm{d}\mu(x)\right)^{1/p}=\|\,\|\mathbf{t}-\mathbf{s}\|_{\mathscr{X}}\,\|_{L_p(\mu)} \tag{2.3}$$

for any pair of measurable functions $\mathbf{t},\mathbf{s}:\mathscr{X}\to\mathscr{X}$. Situations where this inequality holds as equality and $\mathbf{t}$ and $\mathbf{s}$ are optimal maps are related to *compatibility* of the measures μ, $\nu=\mathbf{t}\#\mu$ and $\rho=\mathbf{s}\#\mu$ (see Sect. 2.3.2) and will be of conceptual importance in the context of Fréchet means (see Sect. 3.1).

We also recall the notation $B_R(x_0)=\{x:\|x-x_0\|<R\}$ and $\overline{B}_R(x_0)=\{x:\|x-x_0\|\leq R\}$ for open and closed balls in $\mathscr{X}$.

2.2 Topological Properties

2.2.1 Convergence, Compact Subsets

The topology of a space is determined by the collection of its closed sets. Since $\mathscr{W}_p(\mathscr{X})$ is a metric space, whether a set is closed or not depends on which sequences in $\mathscr{W}_p(\mathscr{X})$ converge. The following characterisation from Villani [124, Theorem 7.12] will be very useful.

Theorem 2.2.1 (Convergence in Wasserstein Space) *Let $\mu,\mu_n\in\mathscr{W}_p(\mathscr{X})$. Then the following are equivalent:*

1. $W_p(\mu_n,\mu)\to 0$ *as* $n\to\infty$;
2. $\mu_n\to\mu$ *weakly and* $\int_{\mathscr{X}}\|x\|^p\,\mathrm{d}\mu_n(x)\to\int_{\mathscr{X}}\|x\|^p\,\mathrm{d}\mu(x)$;
3. $\mu_n\to\mu$ *weakly and*

$$\sup_n\int_{\{x:\|x\|>R\}}\|x\|^p\,\mathrm{d}\mu_n(x)\to 0,\qquad R\to\infty; \tag{2.4}$$

4. *for any* $C>0$ *and any continuous* $f:\mathscr{X}\to\mathbb{R}$ *such that* $|f(x)|\leq C(1+\|x\|^p)$ *for all* x,

$$\int_{\mathscr{X}}f(x)\,\mathrm{d}\mu_n(x)\to\int_{\mathscr{X}}f(x)\,\mathrm{d}\mu(x).$$

5. *(Le Gouic and Loubes [87, Lemma 14]) $\mu_n \to \mu$ weakly and there exists $\nu \in \mathscr{W}_p(\mathscr{X})$ such that $W_p(\mu_n, \nu) \to W_p(\mu, \nu)$.*

Consequently, the Wasserstein topology is finer than the weak topology induced on $\mathscr{W}_p(\mathscr{X})$ from $P(\mathscr{X})$. Indeed, let $\mathscr{A} \subseteq \mathscr{W}_p(\mathscr{X})$ be weakly closed. If $\mu_n \in \mathscr{A}$ converge to μ in $\mathscr{W}_p(\mathscr{X})$, then $\mu_n \to \mu$ weakly, so $\mu \in \mathscr{A}$. In other words, the Wasserstein topology has more closed sets than the induced weak topology. Moreover, each $\mathscr{W}_p(\mathscr{X})$ is a weakly closed subset of $P(\mathscr{X})$ by the same arguments that lead to (1.3). In view of Theorem 2.2.1, a common strategy to establish Wasserstein convergence is to first show tightness and obtain weak convergence, hence a candidate limit, and then show that the stronger Wasserstein convergence actually holds. In some situations, the last part is automatic:

Corollary 2.2.2 *Let $K \subset \mathscr{X}$ be a bounded set and suppose that $\mu_n(K) = 1$ for all $n \geq 1$. Then $W_p(\mu_n, \mu) \to 0$ if and only if $\mu_n \to \mu$ weakly.*

Proof. This is immediate from (2.4).

The fact that convergence in $\mathscr{W}_p$ is stronger than weak convergence is exemplified in the following result. If $\mu_n \to \mu$ and $\nu_n \to \nu$ in $\mathscr{W}_p(\mathscr{X})$, then it is obvious that $W_p(\mu_n, \nu_n) \to W_p(\mu, \nu)$. But if the convergence is only weak, then the Wasserstein distance is still lower semicontinuous:

$$\liminf_{n\to\infty} W_p(\mu_n, \nu_n) \geq W_p(\mu, \nu). \tag{2.5}$$

This follows from Theorem 1.7.2 and (1.3).

Before giving some examples, it will be convenient to formulate Theorem 2.2.1 in probabilistic terms. Let X, X_n be random elements on $\mathscr{X}$ with laws $\mu, \mu_n \in \mathscr{W}_p(\mathscr{X})$. Assume without loss of generality that X, X_n are defined on the same probability space $(\Omega, \mathscr{F}, \mathbb{P})$ and write $W_p(X_n, X)$ to denote $W_p(\mu_n, \mu)$. Then $W_p(X_n, X) \to 0$ if and only if $X_n \to X$ weakly and $\mathbb{E}\|X_n\|^p \to \mathbb{E}\|X\|^p$.

An early example of the use of Wasserstein metric in statistics is due to Bickel and Freedman [21]. Let X_n be independent and identically distributed random variables with mean zero and variance 1 and let Z be a standard normal random variable. Then $Z_n = \sum_{i=1}^n X_i/\sqrt{n}$ converge weakly to Z by the central limit theorem. But $\mathbb{E}Z_n^2 = 1 = \mathbb{E}Z^2$, so $W_2(Z_n, Z) \to 0$. Let Z_n^* be a bootstrapped version of Z_n constructed by resampling the X_n's. If $W_2(Z_n^*, Z_n) \to 0$, then $W_2(Z_n^*, Z) \to 0$ and in particular Z_n^* has the same asymptotic distribution as Z_n.

Another consequence of Theorem 2.2.1 is that (in the presence of weak convergence) convergence of moments automatically yields convergence of smaller moments (there are, however, more elementary ways to see this). In the previous example, for instance, one can also conclude that $\mathbb{E}|Z_n|^p \to \mathbb{E}|Z|^p$ for any $p \leq 2$ by the last condition of the theorem. If in addition $\mathbb{E}X_1^4 < \infty$, then

$$\mathbb{E}Z_n^4 = 3 - \frac{3}{n} + \frac{\mathbb{E}X_1^4}{n} \to 3 = \mathbb{E}Z^4$$

(see Durrett [49, Theorem 2.3.5]) so $W_4(Z_n, Z) \to 0$ and all moments up to order 4 converge.

Condition (2.4) is called *uniform integrability* of the function $x \mapsto \|x\|^p$ with respect to the collection (μ_n). Of course, it holds for a single measure $\mu \in \mathscr{W}_p(\mathscr{X})$ by the dominated convergence theorem. This condition allows us to characterise compact sets in the Wasserstein space. One should beware that when $\mathscr{X}$ is infinite-dimensional, (2.4) alone is not sufficient in order to conclude that μ_n has a convergent subsequence: take μ_n to be Dirac measures at e_n with (e_n) an orthonormal basis of a Hilbert space $\mathscr{X}$ (or any sequence with $\|e_n\| = 1$ that has no convergent subsequence, if $\mathscr{X}$ is a Banach space). The uniform integrability (2.4) must be accompanied with tightness, which is a consequence of (2.4) only when $\mathscr{X} = \mathbb{R}^d$.

Proposition 2.2.3 (Compact Sets in $\mathscr{W}_p$) *A weakly tight set $\mathscr{K} \subseteq \mathscr{W}_p$ is Wasserstein-tight (has a compact closure in $\mathscr{W}_p$) if and only if*

$$\sup_{\mu \in \mathscr{K}} \int_{\{x:\|x\|>R\}} \|x\|^p \, \mathrm{d}\mu(x) \to 0, \qquad R \to \infty. \tag{2.6}$$

Moreover, (2.6) *is equivalent to the existence of a monotonically divergent function $g : \mathbb{R}_+ \to \mathbb{R}_+$ such that*

$$\sup_{\mu \in \mathscr{K}} \int_{\mathscr{X}} \|x\|^p g(\|x\|) \, \mathrm{d}\mu(x) < \infty.$$

The proof is on page 41 of the supplement.

Remark 2.2.4 *For any sequence (μ_n) in $\mathscr{W}_p$ (tight or not) there exists a monotonically divergent g with $\int_{\mathscr{X}} \|x\|^p g(\|x\|) \, \mathrm{d}\mu_n(x) < \infty$ for all n.*

Corollary 2.2.5 (Measures with Common Support) *Let $K \subseteq \mathscr{X}$ be a compact set. Then*

$$\mathscr{K} = \mathscr{W}_p(K) = \{\mu \in P(\mathscr{X}) : \mu(K) = 1\} \subseteq \mathscr{W}_p(\mathscr{X})$$

is compact.

Proof. This is immediate, since $\mathscr{K}$ is weakly tight and the supremum in (2.6) vanishes when R is larger than the finite quantity $\sup_{x \in K} \|x\|$. Finally, K is closed, so $\mathscr{K}$ is weakly closed, hence Wasserstein closed, by the portmanteau Lemma 1.7.1.

For future reference, we give another consequence of uniform integrability, called *uniform absolute continuity*

$$\forall \varepsilon \; \exists \delta \; \forall n \; \forall A \subseteq \mathscr{X} \text{ Borel} : \qquad \mu_n(A) \le \delta \quad \Longrightarrow \quad \int_A \|x\|^p \, \mathrm{d}\mu_n(x) < \varepsilon. \tag{2.7}$$

To show that (2.4) implies (2.7), let $\varepsilon > 0$, choose $R = R_\varepsilon > 0$ such that the supremum in (2.4) is smaller than $\varepsilon/2$, and set $\delta = \varepsilon/(2R^p)$. If $\mu_n(A) \le \delta$, then

$$\int_A \|x\|^p \, \mathrm{d}\mu_n(x) \le \int_{A\cap\overline{B}_R(0)} \|x\|^p \, \mathrm{d}\mu_n(x) + \int_{A\setminus\overline{B}_R(0)} \|x\|^p \, \mathrm{d}\mu_n(x) < \delta R^p + \varepsilon/2 \le \varepsilon.$$

2.2.2 Dense Subsets and Completeness

If we identify a measure $\mu \in \mathcal{W}_p(\mathcal{X})$ with a random variable X (having distribution μ), then X has a finite p-th moment in the sense that the real-valued random variable $\|X\|$ is in L_p. In view of that, it should not come as a surprise that $\mathcal{W}_p(\mathcal{X})$ enjoys topological properties similar to L_p spaces. In this subsection, we give some examples of useful dense subsets of $\mathcal{W}_p(\mathcal{X})$ and then "show" that like $\mathcal{X}$ itself, it is a complete separable metric space. In the next subsection, we describe some of the negative properties that $\mathcal{W}_p(\mathcal{X})$ has, again in similarity with L_p spaces.

We first show that $\mathcal{W}_p(\mathcal{X})$ is separable. The core idea of the proof is the feasibility of approximating any measure with discrete measures as follows.

Let μ be a probability measure on $\mathcal{X}$, and let $X_1, X_2, \dots$ be a sequence of independent random elements in $\mathcal{X}$ with probability distribution μ. Then the *empirical measure* μ_n is defined as the random measure $(1/n)\sum_{i=1}^n \delta\{X_i\}$. The law of large numbers shows that for any (measurable) bounded or nonnegative $f : \mathcal{X} \to \mathbb{R}$, almost surely

$$\int_{\mathcal{X}} f(x) \, \mathrm{d}\mu_n(x) = \frac{1}{n}\sum_{i=1}^n f(X_i) \to \mathbb{E} f(X_1) = \int_{\mathcal{X}} f(x) \, \mathrm{d}\mu(x).$$

In particular when $f(x) = \|x\|^p$, we obtain convergence of moments of order p. Hence by Theorem 2.2.1, if $\mu \in \mathcal{W}_p(\mathcal{X})$, then $\mu_n \to \mu$ in $\mathcal{W}_p(\mathcal{X})$ if and only if $\mu_n \to \mu$ weakly. We know that integrals of bounded functions converge with probability one, but the null set where convergence fails may depend on the chosen function and there are uncountably many such functions. When $\mathcal{X} = \mathbb{R}^d$, by the portmanteau Lemma 1.7.1 we can replace the collection $C_b(\mathcal{X})$ by indicator functions of rectangles of the form $(-\infty, a_1] \times \cdots \times (-\infty, a_d]$ for $a = (a_1, \dots, a_d) \in \mathbb{R}^d$. It turns out that the countable collection provided by rational vectors a suffices (see the proof of Theorem 4.4.1 where this is done in a more complicated setting). For more general spaces $\mathcal{X}$, we need to find another countable collection $\{f_j\}$ such that convergence of the integrals of f_j for all j suffices for weak convergence. Such a collection exists, by using bounded Lipschitz functions (Dudley [47, Theorem 11.4.1]); an alternative construction can be found in Ambrosio et al. [12, Section 5.1]. Thus:

Proposition 2.2.6 (Empirical Measures in $\mathcal{W}_p$) *For any $\mu \in P(\mathcal{X})$ and the corresponding sequence of empirical measures μ_n, $W_p(\mu_n, \mu) \to 0$ almost surely if and only if $\mu \in \mathcal{W}_p(\mathcal{X})$.*

Indeed, if $\mu \notin \mathcal{W}_p(\mathcal{X})$, then $W_p(\mu_n, \mu)$ is infinite for all n, since μ_n is compactly supported, hence in $\mathcal{W}_p(\mathcal{X})$.

Proposition 2.2.6 is the basis for constructing dense subsets of the Wasserstein space.

Theorem 2.2.7 (Dense Subsets of $\mathcal{W}_p$) *The following collections of measures are dense in $\mathcal{W}_p(\mathcal{X})$:*

1. *finitely supported measures with rational weights;*
2. *compactly supported measures;*
3. *finitely supported measures with rational weights on a dense subset $A \subseteq \mathcal{X}$;*
4. *if $\mathcal{X} = \mathbb{R}^d$, the collection of absolutely continuous and compactly supported measures;*
5. *if $\mathcal{X} = \mathbb{R}^d$, the collection of absolutely continuous measures with strictly positive and bounded analytic densities.*

In particular, $\mathcal{W}_p$ is separable (the third set is countable as $\mathcal{X}$ is separable).

This is a simple consequence of Proposition 2.2.6 and approximations, and the details are given on page 43 in the supplement.

Proposition 2.2.8 (Completeness) *The Wasserstein space $\mathcal{W}_p(\mathcal{X})$ is complete.*

One may find two different proofs in Villani [125, Theorem 6.18] and Ambrosio et al. [12, Proposition 7.1.5]. On page 43 of the supplement, we sketch an alternative argument based on completeness of the weak topology.

2.2.3 Negative Topological Properties

In the previous subsection, we have shown that $\mathcal{W}_p(\mathcal{X})$ is separable and complete like L_p spaces. Just like them, however, the Wasserstein space is neither locally compact nor σ-compact. For this reason, existence proofs of Fréchet means in $\mathcal{W}_p(\mathcal{X})$ require tools that are more specific to this space, and do not rely upon local compactness (see Sect. 3.1).

Proposition 2.2.9 ($\mathcal{W}_p$ is Not Locally Compact) *Let $\mu \in \mathcal{W}_p(\mathcal{X})$ and let $\varepsilon > 0$. Then the Wasserstein ball*

$$\overline{B}_\varepsilon(\mu) = \{\nu \in \mathcal{W}_p(\mathcal{X}) : W_p(\mu,\nu) \le \varepsilon\}$$

is not compact.

Ambrosio et al. [12, Remark 7.1.9] show this when μ is a Dirac measure, and we extend their argument on page 43 of the supplement.

From this, we deduce:

Corollary 2.2.10 *The Wasserstein space $\mathcal{W}_p(\mathcal{X})$ is not σ-compact.*

Proof. If $\mathcal{K}$ is a compact set in $\mathcal{W}_p(\mathcal{X})$, then its interior is empty by Proposition 2.2.9. A countable union of compact sets has an empty interior (hence cannot equal the entire space $\mathcal{W}_p(\mathcal{X})$) by the Baire property, which holds on the complete metric space $\mathcal{W}_p(\mathcal{X})$ by the Baire category theorem (Dudley [47, Theorem 2.5.2]).

2.2.4 Covering Numbers

Let $\mathscr{K} \subset \mathscr{W}_p(\mathscr{X})$ be compact and assume that $\mathscr{X} = \mathbb{R}^d$. Then for any $\varepsilon > 0$ the covering number

$$N(\varepsilon;\mathscr{K}) = \min\left\{n : \exists \mu_1,\ldots,\mu_n \in \mathscr{W}_p(\mathscr{X}) \text{ such that } \mathscr{K} \subseteq \bigcup_{i=1}^{n}\{\mu : W_p(\mu,\mu_i) < \varepsilon\}\right\}$$

is finite. These numbers appear in statistics in various ways, particularly in empirical processes (see, for instance, Wainwright [126, Chapter 5]) and the goal of this subsection is to give an upper bound for $N(\varepsilon;\mathscr{K})$. Invoking Proposition 2.2.3, introduce a continuous monotone divergent $f : [0,\infty) \to [0,\infty]$ such that

$$\sup_{\mu\in\mathscr{K}} \int_{\mathbb{R}^d} \|x\|^p f(\|x\|)\, \mathrm{d}\mu(x) \leq 1.$$

The function f provides a certain measure of how compact $\mathscr{K}$ is. If $\mathscr{K} = \mathscr{W}_p(K)$ is the set of measures supported on a compact $K \subseteq \mathbb{R}^d$, then $f(L)$ can be taken infinite for L large, and L can be treated as a constant in the theorem. Otherwise L increases as $\varepsilon \searrow 0$, at a speed that depends on f: the faster f diverges, the slower L grows with decreasing ε and the better the bound becomes.

Theorem 2.2.11 *Let $\varepsilon > 0$ and $L = f^{-1}(1/\varepsilon^p)$. If $d\varepsilon \leq L$, then*

$$\log N(\varepsilon) \leq C_1(d)\left(\frac{L}{\varepsilon}\right)^d \left[(p+d)\log\frac{L}{\varepsilon} + C_2(d,p)\right],$$

with $C_1(d) = 3^d e\theta_d$, $C_2(d,p) = (p+d)\log 3 + (p+2)\log 2 + \log\theta_d$ and $\theta_d = d[5 + \log d + \log\log d]$.

Since $\varepsilon > 0$ is small and L is increasing in ε, the restriction that $d\varepsilon \leq L$ is typically not binding. We provide some examples before giving the proof.

Example 1: if all the measures are supported on the d-dimensional unit ball, then L can be taken equal to one, independently of ε. We obtain

$$\widetilde{N}(\varepsilon) := \frac{\log N(\varepsilon)}{\log 1/\varepsilon} \leq (d+p)C_1(d)\varepsilon^{-d} + \text{ smaller order terms.}$$

Example 2: if all the measures in $\mathscr{K}$ have uniform exponential moments, then $f(L) = e^L$ and $\widetilde{N}(\varepsilon)$ is a constant times $\varepsilon^{-d}[\log 1/\varepsilon]^d$. The exponent p appears only in the constant.

Example 3: suppose that $\mathscr{K}$ is a Wasserstein ball of order $p+\delta$, that is, $f(L) = L^\delta$. Then $L \sim \varepsilon^{-p/\delta}$ and

$$\widetilde{N}(\varepsilon) \leq C_1(d)(p+d)(1+p/\delta)\varepsilon^{-d[1+p/\delta]}$$

up to smaller order terms. Here (when $0 < \delta < \infty$) the behaviour of $\widetilde{N}(\varepsilon)$ depends more strongly upon p: if $p' < p$, then we can replace δ by $\delta' = \delta + p - p' > \delta$, leading to a smaller magnitude of $\widetilde{N}(\varepsilon)$.

Example 4: if $f(L)$ is only $\log L$, then $\widetilde{N}$ behaves like $\varepsilon^{-(d+p)}\exp(\varepsilon^{-pd})$, so p has a very dominant effect.

Proof. The proof is divided into four steps.

Step 1: Compact support. Let $P_L : \mathbb{R}^d \to \mathbb{R}^d$ be the projection onto $\overline{B}_L(0) = \{x \in \mathbb{R}^d : \|x\| \le L\}$ and let $\mu \in \mathscr{K}$. Then

$$W_p^p(\mu, P_L\#\mu) \le \int_{\mathbb{R}^d} \|x - P_L(x)\|^p \,\mathrm{d}\mu(x) = \int_{\|x\|>L} \|x - P_L(x)\|^p \,\mathrm{d}\mu(x)$$
$$\le \int_{\|x\|>L} \|x\|^p \,\mathrm{d}\mu(x) \le \frac{1}{f(L)} \int_{\|x\|>L} \|x\|^p f(\|x\|) \,\mathrm{d}\mu(x) \le \frac{1}{f(L)},$$

and this vanishes as $L \to \infty$.

Step 2: n-Point measures. Let $n = N(\varepsilon; B_L(0))$ be the covering number of the Euclidean ball in $\mathbb{R}^d$. There exists a set $x_1, \dots, x_n \in \mathbb{R}^d$ such that $B_L(0) \subseteq \cup B_\varepsilon(x_i)$. If $\mu \in \mathscr{W}_p(B_L(0))$, there exists a measure μ_n supported on the x_i's and such that

$$W_p(\mu, \mu_n) \le \varepsilon.$$

Indeed let $C_1 = B_\varepsilon(x_1)$, $C_i = B_\varepsilon(x_i) \setminus \cup_{j<i} B_\varepsilon(x_j)$ and define $\mu_n(\{x_i\}) = \mu(C_i)$. The transport map defined by $\mathbf{t}(x) = x_i$ for $x \in C_i$ pushes μ forward to μ_n and

$$W_p^p(\mu_n, \mu) \le \sum_{i=1}^n \int_{C_i} \|x - x_i\|^p \,\mathrm{d}\mu(x) \le \sum_{i=1}^n \varepsilon^p \mu(C_i) = \varepsilon^p.$$

According to Rogers [114], we have the bound

$$n \le e\theta_d [L/\varepsilon]^d, \qquad \theta_d = d[5 + \log d + \log\log d],$$

whenever $\varepsilon \le L/d$.

Step 3: Common weights. If $\mu = \sum a_k \delta_{x_k}$ and $\nu = \sum b_k \delta_{x_k}$, then $W_p^p(\mu, \nu) \le \sum_k |a_k - b_k| \sup_{i,j} \|x_i - x_j\|^p$. Let

$$\mu_{n,\varepsilon,\delta} = \left\{ \sum_{k=1}^n a_k \delta_{x_k} : a_k \in \{0, \delta, 2\delta, \dots, \lceil 1/\delta \rceil \delta\}; \sum a_k = 1 \right\}.$$

This set contains fewer than $(2 + 1/\delta)^{n-1}$ elements, and any measure supported on $\{x_1, \dots, x_n\}$ can be approximated by a measure in $\mu_{n,\varepsilon,\delta}$ with error $2L(n\delta)^{1/p}$.

Step 4: Conclusion. Let $L = f^{-1}(1/\varepsilon^p)$, $n = N(\varepsilon; B_L(0))$ and $\delta = [\varepsilon/(2L)]^p/n$. Combining the previous three steps, we obtain in the case $L \ge \varepsilon d$ that

$$N(3\varepsilon) \le (2 + 1/\delta)^{n-1} \le \left[2 + \left(\frac{L}{\varepsilon}\right)^{p+d} 2^p e\theta_d \right]^{e\theta_d [L/\varepsilon]^d} \le \left[\left(\frac{L}{\varepsilon}\right)^{p+d} 2^{p+2}\theta_d \right]^{e\theta_d [L/\varepsilon]^d},$$

because $L/\varepsilon \geq 1$ and $\theta_d \geq 5$. Conclude that

$$N(\varepsilon) \leq \left[3^{p+d}\left(\frac{L}{\varepsilon}\right)^{p+d} 2^{p+2}\theta_d\right]^{3^d e\theta_d [L/\varepsilon]^d}.$$

2.3 The Tangent Bundle

Although the Wasserstein space $\mathscr{W}_p(\mathscr{X})$ is non-linear in terms of measures, it *is* linear in terms of maps. Indeed, if $\mu \in \mathscr{W}_p(\mathscr{X})$ and $T_i : \mathscr{X} \to \mathscr{X}$ are such that $\|T_i\| \in L_p(\mu)$, then $(\alpha T_1 + \beta T_2)\#\mu \in \mathscr{W}_p(\mathscr{X})$ for all $\alpha, \beta \in \mathbb{R}$. Later, in Sect. 2.4, we shall see that $\mathscr{W}_p(\mathscr{X})$ is in fact homeomorphic to a subset of the space of such functions. The goal of this section is to exploit the linearity of the latter in order to define the tangent bundle of $\mathscr{W}_p$. This in particular will be used for deriving differentiability properties of the Wasserstein distance in Sect. 3.1.6. However, the latter can be understood at a purely analytic level, and readers uncomfortable with differential geometry can access most of the rest of the monograph without reference to this section.

We assume here that $\mathscr{X}$ is a Hilbert space and that $p = 2$; the results extend to any $p > 1$. Absolutely continuous measures are assumed to be so with respect to Lebesgue measure if $\mathscr{X} = \mathbb{R}^d$ and otherwise refer to Definition 1.6.4.

2.3.1 Geodesics, the Log Map and the Exponential Map in $\mathscr{W}_2(\mathscr{X})$

Let $\gamma \in \mathscr{W}_2(\mathscr{X})$ be absolutely continuous and $\mu \in \mathscr{W}_2(\mathscr{X})$ arbitrary. From Sect. 1.6.1, we know that there exists a unique solution to the Monge–Kantorovich problem, and that solution is given by a transport map that we denote by $\mathbf{t}_\gamma^\mu$. Recalling that $\mathbf{i} : \mathscr{X} \to \mathscr{X}$ is the identity map, we can define a curve

$$\gamma_t = \left[\mathbf{i} + t(\mathbf{t}_\gamma^\mu - \mathbf{i})\right]\#\gamma, \qquad t \in [0, 1].$$

This curve is known as McCann's [93] interpolant. As hinted in the introduction to this section, it is constructed via classical linear interpolation of the transport maps $\mathbf{t}_\gamma^\mu$ and the identity. Clearly $\gamma_0 = \gamma$, $\gamma_1 = \mu$ and from (2.3),

$$W_2(\gamma_t, \gamma) \leq \sqrt{\int_{\mathscr{X}} \left[t(\mathbf{t}_\gamma^\mu - \mathbf{i})\right]^2 \mathrm{d}\gamma} \qquad = tW_2(\gamma, \mu);$$

$$W_2(\gamma_t, \mu) \leq \sqrt{\int_{\mathscr{X}} \left[(1-t)(\mathbf{t}_\gamma^\mu - \mathbf{i})\right]^2 \mathrm{d}\gamma} = (1-t)W_2(\gamma, \mu).$$

It follows from the triangle inequality in $\mathscr{W}_2$ that these inequalities must hold as equalities. Taking this one step further, we see that

$$W_2(\gamma_t,\gamma_s)=(t-s)W_2(\gamma,\mu), \qquad 0\le s\le t\le 1.$$

In other words, McCann's interpolant is a *constant-speed geodesic* in $\mathcal{W}_2(\mathcal{X})$.

In view of this, it seems reasonable to define the *tangent space* of $\mathcal{W}_2(\mathcal{X})$ at μ as (Ambrosio et al. [12, Definition 8.5.1])

$$\mathrm{Tan}_\mu=\overline{\{t(\mathbf{t}-\mathbf{i}):\mathbf{t}=\mathbf{t}_\mu^\nu \text{ for some } \nu\in\mathcal{W}_2(\mathcal{X});t>0\}}^{L_2(\mu)}.$$

It follows from the definition that $\mathrm{Tan}_\mu\subseteq L_2(\mu)$. (Strictly speaking, Tan_μ is a subset of the space of functions $f:\mathcal{X}\to\mathcal{X}$ such that $\|f\|\in L_2(\mu)$ rather than $L_2(\mu)$ itself, as in Definition 2.4.3, but we will write L_2 for simplicity.)

Although not obvious from the definition, this is a linear space. The reason is that, in $\mathbb{R}^d$, Lipschitz functions are dense in $L_2(\mu)$, and for $\mathbf{t}$ Lipschitz the negative of a tangent element

$$-t(\mathbf{t}-\mathbf{i})=s(\mathbf{s}-\mathbf{i}), \qquad s>t\|\mathbf{t}\|_{\mathrm{Lip}}, \qquad \mathbf{s}=\mathbf{i}+\frac{t}{s}(\mathbf{i}-\mathbf{t})$$

lies in the tangent space, since $\mathbf{s}$ can be seen to belong to the subgradient of a convex function by definition of s. This also shows that Tan_μ can be seen to be the $L_2(\mu)$-closure of all gradients of C_c^∞ functions. We refer to [12, Definition 8.4.1 and Theorem 8.5.1] for the proof and extensions to other values of $p>1$ and to infinite dimensions, using cylindrical functions that depend on finitely many coordinates [12, Definition 5.1.11]. The alternative definition highlights that it is essentially the inner product in Tan_μ, but not the elements themselves, that depends on μ.

The tangent space definition is valid for arbitrary measures in $\mathcal{W}_2(\mathcal{X})$. The exponential map at $\gamma\in\mathcal{W}_2(\mathcal{X})$ is the restriction to Tan_γ of the mapping that sends $\mathbf{r}\in L_2(\gamma)$ to $[\mathbf{r}+\mathbf{i}]\#\gamma\in\mathcal{W}_2(\mathcal{X})$. More explicitly, $\exp_\gamma:\mathrm{Tan}_\gamma\to\mathcal{W}_2$ takes the form

$$\exp_\gamma(t(\mathbf{t}-\mathbf{i}))=\exp_\gamma([t\mathbf{t}+(1-t)\mathbf{i}]-\mathbf{i})=[t\mathbf{t}+(1-t)\mathbf{i}]\#\gamma \quad (t\in\mathbb{R}).$$

Thus, when γ is absolutely continuous, $\exp_\gamma$ is surjective, as can be seen from its right inverse, the log map

$$\log_\gamma:\mathcal{W}_2\to\mathrm{Tan}_\gamma \qquad \log_\gamma(\mu)=\mathbf{t}_\gamma^\mu-\mathbf{i},$$

defined throughout $\mathcal{W}_2$ (by virtue of Theorem 1.6.2). In symbols,

$$\exp_\gamma(\log_\gamma(\mu))=\mu, \quad \mu\in\mathcal{W}_2, \quad \text{and} \quad \log_\gamma(\exp_\gamma(t(\mathbf{t}-\mathbf{i})))=t(\mathbf{t}-\mathbf{i}) \quad (t\in[0,1]),$$

because convex combinations of optimal maps are optimal maps as well. In particular, McCann's interpolant $\left[\mathbf{i}+t(\mathbf{t}_\gamma^\mu-\mathbf{i})\right]\#\gamma$ is mapped bijectively to the line segment $t(\mathbf{t}_\gamma^\mu-\mathbf{i})\in\mathrm{Tan}_\gamma$ through the log map.

It is also worth mentioning that McCann's interpolant can also be defined as

$$[tp_2+(1-t)p_1]\#\pi, \qquad p_1(x,y)=x, \quad p_2(x,y)=y,$$

where $p_1, p_2 : \mathscr{X}^2 \to \mathscr{X}$ are projections and π is any optimal transport plan between γ and μ. This is defined for arbitrary measures $\gamma, \mu \in \mathscr{W}_2$, and reduces to the previous definition if γ is absolutely continuous. It is shown in Ambrosio et al. [12, Chapter 7] or Santambrogio [119, Proposition 5.32] that these are the only constant-speed geodesics in $\mathscr{W}_2$.

2.3.2 Curvature and Compatibility of Measures

Let $\gamma, \mu, \nu \in \mathscr{W}_2(\mathscr{X})$ be absolutely continuous measures. Then by (2.3)

$$W_2^2(\mu,\nu) \le \int_{\mathscr{X}} \|\mathbf{t}_\gamma^\mu(x) - \mathbf{t}_\gamma^\nu(x)\|^2 \, d\gamma(x) = \|\log_\gamma(\mu) - \log_\gamma(\nu)\|^2.$$

In other words, the distance between μ and ν is smaller in $\mathscr{W}_2(\mathscr{X})$ than the distance between the corresponding vectors $\log_\gamma(\mu)$ and $\log_\gamma(\nu)$ in the tangent space Tan_γ. In the terminology of differential geometry, this means that the Wasserstein space has *nonnegative sectional curvature* at any absolutely continuous γ.

It is instructive to see when equality holds. As $\mathbf{t}_\nu^\gamma = (\mathbf{t}_\gamma^\nu)^{-1}$, a change of variables gives

$$W_2^2(\mu,\nu) \le \int_{\mathscr{X}} \|\mathbf{t}_\gamma^\mu(\mathbf{t}_\nu^\gamma(x)) - x\|^2 \, d\nu(x).$$

Since the map $\mathbf{t}_\gamma^\mu \circ \mathbf{t}_\nu^\gamma$ pushes forward ν to μ, equality holds if and only if $\mathbf{t}_\gamma^\mu \circ \mathbf{t}_\nu^\gamma = \mathbf{t}_\nu^\mu$. This motivates the following definition.

Definition 2.3.1 (Compatible Measures) *A collection of absolutely continuous measures $\mathscr{C} \subseteq \mathscr{W}_2(\mathscr{X})$ is* compatible *if for all $\gamma, \mu, \nu \in \mathscr{C}$, we have $\mathbf{t}_\gamma^\mu \circ \mathbf{t}_\nu^\gamma = \mathbf{t}_\nu^\mu$ (in $L_2(\nu)$).*

Remark 2.3.2 *The absolute continuity is not necessary and was introduced for notational simplicity. A more general definition that applies to general measures is the following: every finite subcollection of $\mathscr{C}$ admits an optimal multicoupling whose relevant projections are simultaneously pairwise optimal; see the paragraph preceding Theorem 3.1.9.*

A collection of two (absolutely continuous) measures is always compatible. More interestingly, if $\mathscr{X} = \mathbb{R}$, then the entire collection of absolutely continuous (or even just continuous) measures is compatible. This is because of the simple geometry of convex functions in $\mathbb{R}$: gradients of convex functions are nondecreasing, and this property is stable under composition. In a more probabilistic way of thinking, one can always push-forward μ to ν via the uniform distribution $\mathrm{Leb}|_{[0,1]}$ (see Sect. 1.5). Letting F_μ^{-1} and F_ν^{-1} denote the quantile functions, we have seen that

$$W_2(\mu,\nu) = \|F_\mu^{-1} - F_\nu^{-1}\|_{L_2(0,1)}.$$

(As a matter of fact, in this specific case, the equality holds for all $p \geq 1$ and not only for $p = 2$.) In other words, $\mu \mapsto F_\mu^{-1}$ is an *isometry* from $\mathscr{W}_2(\mathbb{R})$ to the subset of $L_2(0,1)$ formed by (equivalence classes of) left-continuous nondecreasing functions on $(0,1)$. Since this is a convex subset of a Hilbert space, this property provides a very simple way to evaluate Fréchet means in $\mathscr{W}_2(\mathbb{R})$ (see Sect. 3.1). If $\gamma = \mathrm{Leb}|_{[0,1]}$, then $F_\mu^{-1} = \mathbf{t}_\gamma^\mu$ for all μ, so we can write the above equality as

$$W_2^2(\mu,\nu) = \|F_\mu^{-1} - F_\nu^{-1}\|_{L_2(0,1)} = \|\log_\gamma(\mu) - \log_\gamma(\nu)\|^2,$$

so that if $\mathscr{X} = \mathbb{R}$, the Wasserstein space is essentially *flat* (has zero sectional curvature).

The importance of compatibility can be seen as mimicking the simple one-dimensional case in terms of a Hilbert space embedding. Let $\mathscr{C} \subseteq \mathscr{W}_2(\mathscr{X})$ be compatible and fix $\gamma \in \mathscr{C}$. Then for all $\mu, \nu \in \mathscr{C}$

$$W_2^2(\mu,\nu) = \int_{\mathscr{X}} \|\mathbf{t}_\gamma^\mu(x) - \mathbf{t}_\gamma^\nu(x)\|^2 \, \mathrm{d}\gamma(x) = \|\log_\gamma(\mu) - \log_\gamma(\nu)\|_{L_2(\gamma)}^2.$$

Consequently, once again, $\mu \mapsto \mathbf{t}_\gamma^\mu$ is an isometric embedding of $\mathscr{C}$ into $L_2(\gamma)$. Generalising the one-dimensional case, we shall see that this allows for easy calculations of Fréchet means by means of averaging transport maps (Theorem 3.1.9).

Example: Gaussian compatible measures. The Gaussian case presented in Sect. 1.6.3 is helpful in shedding light on the structure imposed by the compatibility condition. Let $\gamma \in \mathscr{W}_2(\mathbb{R}^d)$ be a standard Gaussian distribution with identity covariance matrix. Let Σ_μ denote the covariance matrix of a measure $\mu \in \mathscr{W}_2(\mathbb{R}^d)$. When μ and ν are centred nondegenerate Gaussian measures,

$$\mathbf{t}_\gamma^\mu = \Sigma_\mu^{1/2}; \qquad \mathbf{t}_\gamma^\nu = \Sigma_\nu^{1/2}; \qquad \mathbf{t}_\mu^\nu = \Sigma_\mu^{-1/2}[\Sigma_\mu^{1/2}\Sigma_\nu\Sigma_\mu^{1/2}]^{1/2}\Sigma_\mu^{-1/2},$$

so that γ, μ, and ν are compatible if and only if

$$\mathbf{t}_\mu^\nu = \mathbf{t}_\gamma^\nu \circ \mathbf{t}_\mu^\gamma = \Sigma_\nu^{1/2}\Sigma_\mu^{-1/2}.$$

Since the matrix on the left-hand side must be symmetric, it must necessarily be that $\Sigma_\nu^{1/2}$ and $\Sigma_\mu^{-1/2}$ commute (if A and B are symmetric, then AB is symmetric if and only if $AB = BA$), or equivalently, if and only if Σ_ν and Σ_μ commute. We see that a collection $\mathscr{C}$ of Gaussian measures on $\mathbb{R}^d$ that includes the standard Gaussian distribution is compatible if and only if all the covariance matrices of the measures in $\mathscr{C}$ are *simultaneously diagonalisable*. In other words, there exists an orthogonal matrix U such that $D_\mu = U\Sigma_\mu U^t$ is diagonal for all $\mu \in \mathscr{C}$. In that case, formula (1.6)

$$\mathscr{W}_2^2(\mu,\nu) = \mathrm{tr}[\Sigma_\mu + \Sigma_\nu - 2(\Sigma_\mu^{1/2}\Sigma_\nu\Sigma_\mu^{1/2})^{1/2}] = \mathrm{tr}[\Sigma_\mu + \Sigma_\nu - 2\Sigma_\mu^{1/2}\Sigma_\nu^{1/2}]$$

simplifies to

$$\mathscr{W}_2^2(\mu,\nu)=\mathrm{tr}[D_\mu+D_\nu-2D_\mu^{1/2}D_\nu^{1/2}]=\sum_{i=1}^d(\sqrt{\alpha_i}-\sqrt{\beta_i})^2,\quad \alpha_i=[D_\mu]_{ii};\ \beta_i=[D_\nu]_{ii},$$

and identifying the (nonnegative) number $a\in\mathbb{R}$ with the map $x\mapsto ax$ on $\mathbb{R}$, the optimal maps take the "orthogonal separable" form

$$\mathbf{t}_\mu^\nu=\Sigma_\nu^{1/2}\Sigma_\mu^{-1/2}=UD_\nu^{1/2}D_\mu^{-1/2}U^t=U\circ\left(\sqrt{\beta_1/\alpha_1},\dots,\sqrt{\beta_d/\alpha_d}\right)\circ U^t.$$

In other words, up to an orthogonal change of coordinates, the optimal maps take the form of d nondecreasing real-valued functions. This is yet another crystallisation of the one-dimensional-like structure of compatible measures.

With the intuition of the Gaussian case at our disposal, we can discuss a more general case. Suppose that the optimal maps are continuously differentiable. Then differentiating the equation $\mathbf{t}_\mu^\nu=\mathbf{t}_\gamma^\nu\circ\mathbf{t}_\mu^\gamma$ gives

$$\nabla\mathbf{t}_\mu^\nu(x)=\nabla\mathbf{t}_\gamma^\nu(\mathbf{t}_\mu^\gamma(x))\nabla\mathbf{t}_\mu^\gamma(x).$$

Since optimal maps are gradients of convex functions, their derivatives must be symmetric and positive semidefinite matrices. A product of such matrices stays symmetric if and only if they commute, so in this differentiable setting, compatibility is equivalent to commutativity of the matrices $\nabla\mathbf{t}_\gamma^\nu(\mathbf{t}_\mu^\gamma(x))$ and $\nabla\mathbf{t}_\mu^\gamma(x)$ for μ-almost all x. In the Gaussian case, the optimal maps are linear functions, so x does not appear in the matrices.

Here are some examples of compatible measures. It will be convenient to describe them using the optimal maps from a reference measure $\gamma\in\mathscr{W}_2(\mathbb{R}^d)$. Define $\mathscr{C}=\mathbf{t}\#\gamma$ with $\mathbf{t}$ belonging to one of the following families. The first imposes the one-dimensional structure by varying only the behaviour of the norm of x, while the second allows for separation of variables that splits the d-dimensional problem into d one-dimensional ones.

Radial transformations. Consider the collection of functions $\mathbf{t}:\mathbb{R}^d\to\mathbb{R}^d$ of the form $\mathbf{t}(x)=xG(\|x\|)$ with $G:\mathbb{R}_+\to\mathbb{R}$ differentiable. Then a straightforward calculation shows that

$$\nabla\mathbf{t}(x)=G(\|x\|)I+[G'(\|x\|)/\|x\|]\,xx^t.$$

Since both I and xx^t are positive semidefinite, the above matrix is so if both G and G' are nonnegative. If $\mathbf{s}(x)=xH(\|x\|)$ is a function of the same form, then $\mathbf{s}(\mathbf{t}(x))=xG(\|x\|)H(\|x\|G(\|x\|))$ which belongs to that family of functions (since G is nonnegative). Clearly

$$\nabla\mathbf{s}(\mathbf{t}(x))=H\left[\|x\|G(\|x\|)\right]I+\left[G(\|x\|)H'(\|x\|G(\|x\|))/\|x\|\right]xx^t$$

commutes with $\nabla\mathbf{t}(x)$, since both matrices are of the form $aI+bxx^t$ with a,b scalars (that depend on x). In order to be able to change the base measure γ, we need to check that the inverses belong to the family. But if $y = \mathbf{t}(x)$, then $x = ay$ for some scalar a that solves the equation

$$aG(a\|y\|) = 1.$$

Such a is guaranteed to be unique if $a \mapsto aG(a)$ is strictly increasing and it will exist (for y in the range of $\mathbf{t}$) if it is continuous. As a matter of fact, since the eigenvalues of $\nabla \mathbf{t}(x)$ are $G(a)$ and

$$G(a) + G'(a)a = (aG(a))', \qquad a = \|x\|,$$

the condition that $a \mapsto aG(a)$ is strictly increasing is sufficient (this is weaker than G itself increasing). Finally, differentiability of G is not required, so it is enough if G is continuous and $aG(a)$ is strictly increasing.

Separable variables. Consider the collection of functions $\mathbf{t} : \mathbb{R}^d \to \mathbb{R}^d$ of the form

$$\mathbf{t}(x_1, \dots, x_d) = (T_1(x_1), \dots, T_d(x_d)), \qquad T_i : \mathbb{R} \to \mathbb{R}, \tag{2.8}$$

with T_i continuous and strictly increasing. This is a generalisation of the compatible Gaussian case discussed above in which all the T_i's were linear. Here, it is obvious that elements in this family are optimal maps and that the family is closed under inverses and composition, so that compatibility follows immediately.

This family is characterised by measures having a *common dependence structure*. More precisely, we say that $C : [0,1]^d \to [0,1]$ is a *copula* if C is (the restriction of) a distribution function of a random vector having uniform margins. In other words, if there is a random vector $V = (V_1, \dots, V_d)$ with $\mathbb{P}(V_i \le a) = a$ for all $a \in [0,1]$ and all $j = 1, \dots, d$, and

$$\mathbb{P}(V_1 \le v_1, \dots, V_d \le v_d) = C(v_1, \dots, v_d), \qquad u_i \in [0,1].$$

Nelsen [97] provides an overview on copulae. To any d-dimensional probability measure μ, one can assign a copula $C = C_\mu$ in terms of the distribution function G of μ and its marginals G_j as

$$G(a_1, \dots, a_d) = \mu((-\infty, a_1] \times \cdots \times (-\infty, a_d]) = C(G_1(a_1), \dots, G_d(a_d)).$$

If each G_j is surjective on $(0,1)$, which is equivalent to it being continuous, then this equation defines C uniquely on $(0,1)^d$, and consequently on $[0,1]^d$. If some marginal G_j is not continuous, then uniqueness is lost, but C still exists [97, Chapter 2]. The connection of copulae to compatibility becomes clear in the following lemma, proven on page 51 in the supplement.

Lemma 2.3.3 (Compatibility and Copulae) *The copulae associated with absolutely continuous measures $\mu, \nu \in \mathscr{W}_2(\mathbb{R}^d)$ are equal if and only if $\mathbf{t}_\mu^\nu$ takes the separable form* (2.8).

Composition with linear functions. If $\phi : \mathbb{R}^d \to \mathbb{R}$ is convex with gradient $\mathbf{t}$ and A is a $d \times d$ matrix, then the gradient of the convex function $x \mapsto \phi(Ax)$ at x is $\mathbf{t}_A = A^t \mathbf{t}(Ax)$.

Suppose ψ is another convex function with gradient $\mathbf{s}$ and that compatibility holds, i.e., $\nabla\mathbf{s}(\mathbf{t}(x))$ commutes with $\nabla\mathbf{t}(x)$ for all x. Then in order for

$$\nabla\mathbf{s}_A(\mathbf{t}_A(x)) = A^t\nabla\mathbf{s}(AA^t\mathbf{t}(Ax))A \qquad \text{and} \qquad \nabla\mathbf{t}_A(x) = A^t\nabla\mathbf{t}(Ax)A$$

to commute, it suffices that $AA^t = I$, i.e., that A be orthogonal. Consequently, if $\{\mathbf{t}\#\mu\}_{\mathbf{t}\in\mathbf{T}}$ are compatible, then so are $\{\mathbf{t}_U\#\mu\}_{\mathbf{t}\in\mathbf{T}}$ for any orthogonal matrix U.

2.4 Random Measures in Wasserstein Space

Let μ be a fixed absolutely continuous probability measure in $\mathcal{W}_2(\mathcal{X})$. If $\Lambda \in \mathcal{W}_2(\mathcal{X})$ is another probability measure, then the transport map $\mathbf{t}_\mu^\Lambda$ and the convex potential are functions of Λ. If Λ is now random, then we would like to be able to make probability statements about them. To this end, it needs to be shown that $\mathbf{t}_\mu^\Lambda$ and the convex potential are *measurable* functions of Λ. The goal of this section is to develop a rigorous mathematical framework that justifies such probability statements. We show that all the relevant quantities are indeed measurable, and in particular establish a Fubini-type result in Proposition 2.4.9. This technical section may be skipped at first reading.

Here is an example of a measurability result (Villani [125, Corollary 5.22]). Recall that $P(\mathcal{X})$ is the space of Borel probability measures on $\mathcal{X}$, endowed with the topology of weak convergence that makes it a metric space. Let $\mathcal{X}$ be a complete separable metric space and $c: \mathcal{X}^2 \to \mathbb{R}_+$ a continuous cost function. Let $(\Omega, \mathcal{F}, \mathbb{P})$ be a probability space and $\Lambda, \kappa: \Omega \to P(\mathcal{X})$ be measurable maps. Then there exists a *measurable selection* of optimal transference plans. That is, a measurable $\pi: \Omega \to P(\mathcal{X}^2)$ such that $\pi(\omega) \in \Pi(\Lambda(\omega), \kappa(\omega))$ is optimal for all $\omega \in \Omega$.

Although this result is very general, it only provides information about π. If π is induced from a map T, it is not obvious how to construct T from π in a measurable way; we will therefore follow a different path. In order to (almost) have a self-contained exposition, we work in a somewhat simplified setting that nevertheless suffices for the sequel. At least in the Euclidean case $\mathcal{X} = \mathbb{R}^d$, more general measurability results in the flavour of this section can be found in Fontbona et al. [53]. On the other hand, we will not need to appeal to abstract measurable selection theorems as in [53, 125].

2.4.1 Measurability of Measures and of Optimal Maps

Let $\mathcal{X}$ be a separable Banach space. (Most of the results below hold for any complete separable metric space but we will avoid this generality for brevity and simpler notation). The Wasserstein space $\mathcal{W}_p(\mathcal{X})$ is a metric space for any $p \geq 1$. We can thus define:

Definition 2.4.1 (Random Measure) *A random measure Λ is any measurable map from a probability space $(\Omega, \mathscr{F}, \mathbb{P})$ to $\mathscr{W}_p(\mathscr{X})$, endowed with its Borel σ-algebra.*

In what follows, whenever we call something random, we mean that it is measurable as a map from some generic unspecified probability space.

Lemma 2.4.2 *A random measure Λ is measurable if and only if it is measurable with respect to the induced weak topology.*

Since both topologies are Polish, this follows from abstract measure-theoretic results (Fremlin [57, Paragraph 423F]). We give an elementary proof on page 53 of the supplement.

Optimal maps are functions from $\mathscr{X}$ to itself. In order to define random optimal maps, we need to define a topology and a σ-algebra on the space of such functions.

Definition 2.4.3 (The Space $\mathscr{L}_p(\mu)$) *Let $\mathscr{X}$ be a Banach space and μ a Borel measure on $\mathscr{X}$. Then the space $\mathscr{L}_p(\mu)$ is the space of measurable functions $f : \mathscr{X} \to \mathscr{X}$ such that*

$$\|f\|_{\mathscr{L}_p(\mu)} = \left(\int_{\mathscr{X}} \|f(x)\|_{\mathscr{X}}^p \, \mathrm{d}\mu(x) \right)^{1/p} < \infty.$$

When $\mathscr{X}$ is separable, $\mathscr{L}_p(\mu)$ is an example of a *Bochner space*, though we will not use this terminology.

It follows from the definition that $\|f\|_{\mathscr{L}_p(\mu)}$ is the L_p norm of the map $x \mapsto \|f(x)\|_{\mathscr{X}}$ from $\mathscr{X}$ to $\mathbb{R}$:

$$\|f\|_{\mathscr{L}_p(\mu)} = \| \; \|f\|_{\mathscr{X}} \; \|_{L_p(\mu)}.$$

As usual we identify functions that coincide almost everywhere. Clearly, $\mathscr{L}_p(\mu)$ is a normed vector space. It enjoys another property shared by L_p spaces—completeness:

Theorem 2.4.4 (Riesz–Fischer) *The space $\mathscr{L}_p(\mu)$ is a Banach space.*

The proof, a simple variant of the classical one, is given on page 53 of the supplement.

Random maps lead naturally to random measures:

Lemma 2.4.5 (Push-Forward with Random Maps) *Let $\mu \in \mathscr{W}_p(\mathscr{X})$ and let $\mathbf{t}$ be a random map in $\mathscr{L}_p(\mu)$. Then $\Lambda = \mathbf{t}\#\mu$ is a continuous mapping from $\mathscr{L}_p(\mu)$ to $\mathscr{W}_p(\mathscr{X})$, hence a random measure.*

Proof. That Λ takes values in $\mathscr{W}_p$ follows from a change of variables

$$\int_{\mathscr{X}} \|x\|^p \, \mathrm{d}\Lambda(x) = \int_{\mathscr{X}} \|\mathbf{t}(x)\|^p \, \mathrm{d}\mu(x) = \|\mathbf{t}\|_{\mathscr{L}_p(\mu)} < \infty.$$

Since $W_p(\mathbf{t}\#\mu, \mathbf{s}\#\mu) \le \| \; \|\mathbf{t} - \mathbf{s}\|_{\mathscr{X}} \; \|_{L_p(\mu)} = \|\mathbf{t} - \mathbf{s}\|_{\mathscr{L}_p(\mu)}$ (see (2.3)), Λ is a continuous (in fact, 1-Lipschitz) function of $\mathbf{t}$.

Conversely, $\mathbf{t}$ is a continuous function of Λ:

Lemma 2.4.6 (Measurability of Transport Maps) *Let Λ be a random measure in $\mathscr{W}_p(\mathscr{X})$ and let $\mu \in \mathscr{W}_p(\mathscr{X})$ such that $(\mathbf{i}, \mathbf{t}_\mu^\Lambda)\#\mu$ is the unique optimal coupling of μ and Λ. Then $\Lambda \mapsto \mathbf{t}_\mu^\Lambda$ is a continuous mapping from $\mathscr{W}_p(\mathscr{X})$ to $\mathscr{L}_p(\mu)$, so $\mathbf{t}_\mu^\Lambda$ is a random element in $\mathscr{L}_p(\mu)$. In particular, the result holds if $\mathscr{X}$ is a separable Hilbert space, $p > 1$, and μ is absolutely continuous.*

Proof. This result is more subtle than Lemma 2.4.5, since $\Lambda \mapsto \mathbf{t}_\mu^\Lambda$ is not necessarily Lipschitz. We give here a self-contained proof for the Euclidean case with quadratic cost and μ absolutely continuous. The general case builds on Villani [125, Corollary 5.23] and is given on page 54 of the supplement.

Suppose that $\Lambda_n \to \Lambda$ in $\mathscr{W}_2(\mathbb{R}^d)$ and fix $\varepsilon > 0$. For any $S \subseteq \mathbb{R}^d$,

$$\|\mathbf{t}_\mu^{\Lambda_n} - \mathbf{t}_\mu^\Lambda\|_{\mathscr{L}_2(\mu)}^2 = \int_S \|\mathbf{t}_\mu^{\Lambda_n} - \mathbf{t}_\mu^\Lambda\|^2 \, d\mu + \int_{\mathbb{R}^d \setminus S} \|\mathbf{t}_\mu^{\Lambda_n} - \mathbf{t}_\mu^\Lambda\|^2 \, d\mu.$$

Since $\|a - b\|^p \le 2^p \|a\|^p + 2^p \|b\|^p$, the last integral is no larger than

$$4 \int_{\mathbb{R}^d \setminus S} \|\mathbf{t}_\mu^{\Lambda_n}\|^2 \, d\mu + 4 \int_{\mathbb{R}^d \setminus S} \|\mathbf{t}_\mu^\Lambda\|^2 \, d\mu = 4 \int_{(\mathbf{t}_\mu^{\Lambda_n})^{-1}(\mathbb{R}^d \setminus S)} \|x\|^2 \, d\Lambda_n(x) + 4 \int_{(\mathbf{t}_\mu^\Lambda)^{-1}(\mathbb{R}^d \setminus S)} \|x\|^2 \, d\Lambda(x).$$

Since (Λ_n) and Λ are tight in the Wasserstein space, they must satisfy the absolute uniform continuity (2.7). Let $\delta = \delta_\varepsilon$ as in (2.7), and notice that by the measure preserving property of the optimal maps, the last two integrals are taken on sets of measures $1 - \mu(S)$. Since μ is absolutely continuous, we can find a compact set S of μ-measure at least $1 - \delta$ and on which Proposition 1.7.11 applies (see Corollary 1.7.12), yielding

$$\int_S \|\mathbf{t}_\mu^{\Lambda_n} - \mathbf{t}_\mu^\Lambda\|^2 \, d\mu \le \|\mathbf{t}_\mu^{\Lambda_n} - \mathbf{t}_\mu^\Lambda\|_\infty^2 \to 0, \qquad n \to \infty,$$

so that

$$\limsup_{n \to \infty} \|\mathbf{t}_\mu^{\Lambda_n} - \mathbf{t}_\mu^\Lambda\|_{\mathscr{L}_2(\mu)} \le 8\varepsilon,$$

and this completes the proof upon letting $\varepsilon \to 0$.

In Proposition 5.3.7, we show under some conditions that $\|\mathbf{t}_\mu^\Lambda\|_{\mathscr{L}_2(\mu)}$ is a continuous function of μ.

2.4.2 Random Optimal Maps and Fubini's Theorem

From now on, we assume that $\mathscr{X}$ is a separable Hilbert space and that $p = 2$. The results can most likely be generalised to all $p > 1$ (see Ambrosio et al. [12, Section 10.2]), but we restrict to the quadratic case for simplicity.

Theorem 3.2.13 below requires the application of Fubini's theorem in the form

$$\mathbb{E}\int_{\mathscr{X}}\left\langle \mathbf{t}_{\theta_0}^{\Lambda}-\mathbf{i},\mathbf{t}_{\theta_0}^{\theta}-\mathbf{i}\right\rangle \mathrm{d}\theta_0=\int_{\mathscr{X}}\mathbb{E}\left\langle \mathbf{t}_{\theta_0}^{\Lambda}-\mathbf{i},\mathbf{t}_{\theta_0}^{\theta}-\mathbf{i}\right\rangle \mathrm{d}\theta_0=\int_{\mathscr{X}}\left\langle \mathbb{E}\mathbf{t}_{\theta_0}^{\Lambda}-\mathbf{i},\mathbf{t}_{\theta_0}^{\theta}-\mathbf{i}\right\rangle \mathrm{d}\theta_0.$$

In order for this to even make sense, we need to have a meaning for "expectation" in the spaces $\mathscr{L}_2(\theta_0)$ and $L_2(\theta_0)$, both of which are Banach spaces. There are several (nonequivalent) definitions for integrals in such spaces (Hildebrant [69]); the one which will be the most convenient for our needs is the Bochner integral.

Definition 2.4.7 (Bochner Integral) *Let B be a Banach space and let $f:(\Omega,\mathscr{F},\mathbb{P})\to B$ be a simple random element taking values in B:*

$$f(\omega)=\sum_{j=1}^{n}f_j\mathbf{1}\{\omega\in\Omega_j\},\qquad \Omega_j\in\mathscr{F},\qquad f_j\in B.$$

Then the Bochner integral (or expectation) of f is defined by

$$\mathbb{E}f=\sum_{j=1}^{n}\mathbb{P}(\Omega_j)f_j\in B.$$

If f is measurable and there exists a sequence f_n of simple random elements such that $\|f_n-f\|\to 0$ almost surely and $\mathbb{E}\|f_n-f\|\to 0$, then the Bochner integral of f is defined as the limit

$$\mathbb{E}f=\lim_{n\to\infty}\mathbb{E}f_n.$$

The space of functions for which the Bochner integral is defined is the *Bochner space* $L_1(\Omega;B)$, but we will use neither this terminology nor the notation. It is not difficult to see that Bochner integrals are well-defined: the expectations do not depend on the representation of the simple functions nor on the approximating sequence, and the limit exists in B (because it is complete). More on Bochner integrals can be found in Hsing and Eubank [71, Section 2.6] or Dunford et al. [48, Chapter III.6]. A major difference from the real case is that there is no clear notion of "infinity" here: the Bochner integral is always an element of B, whereas expectations of real-valued random variables can be defined in $\mathbb{R}\cup\{\pm\infty\}$. It turns out that separability is quite important in this setting:

Lemma 2.4.8 (Approximation of Separable Functions) *Let $f:\Omega\to B$ be measurable. Then there exists a sequence of simple functions f_n such that $\|f_n(\omega)-f(\omega)\|\to 0$ for almost all ω if and only if $f(\Omega\setminus\mathscr{N})$ is separable for some $\mathscr{N}\subseteq\Omega$ of probability zero. In that case, f_n can be chosen so that $\|f_n(\omega)\|\leq 2\|f(\omega)\|$ for all $\omega\in\Omega$.*

A proof can be found in [48, Lemma III.6.9], or on page 55 of the supplement. Functions satisfying this approximation condition are sometimes called *strongly measurable* or *Bochner measurable*. In view of the lemma, we will call them *separately valued*, since this is the condition that will need to be checked in order to define their integrals.

Two remarks are in order. Firstly, if B itself is separable, then $f(\Omega)$ will obviously be separable. Secondly, the set $\mathcal{N}' \subset \Omega \setminus \mathcal{N}$ on which (g_{n_k}) does not converge to f may fail to be measurable, but must have outer probability zero (it is included in a measurable set of measure zero) [48, Lemma III.6.9]. This can be remedied by assuming that the probability space $(\Omega, \mathscr{F}, \mathbb{P})$ is complete. It will not, however, be necessary to do so, since this measurability issue will not alter the Bochner expectation of f.

Proposition 2.4.9 (Fubini for Optimal Maps) *Let Λ be a random measure in $\mathscr{W}_2(\mathscr{X})$ such that $\mathbb{E}W_2(\delta_0, \Lambda) < \infty$ and let $\theta_0, \theta \in \mathscr{W}_2(\mathscr{X})$ such that $\mathbf{t}_{\theta_0}^{\Lambda}$ and $\mathbf{t}_{\theta_0}^{\theta}$ exist (and are unique) with probability one. (For example, if θ_0 is absolutely continuous.) Then*

$$\mathbb{E}\int_{\mathscr{X}} \left\langle \mathbf{t}_{\theta_0}^{\Lambda} - \mathbf{i}, \mathbf{t}_{\theta_0}^{\theta} - \mathbf{i} \right\rangle \mathrm{d}\theta_0 = \int_{\mathscr{X}} \mathbb{E}\left\langle \mathbf{t}_{\theta_0}^{\Lambda} - \mathbf{i}, \mathbf{t}_{\theta_0}^{\theta} - \mathbf{i} \right\rangle \mathrm{d}\theta_0 = \int_{\mathscr{X}} \left\langle \mathbb{E}\mathbf{t}_{\theta_0}^{\Lambda} - \mathbf{i}, \mathbf{t}_{\theta_0}^{\theta} - \mathbf{i} \right\rangle \mathrm{d}\theta_0. \tag{2.9}$$

This holds by linearity when Λ is a simple random measure. The general case follows by approximation: the Wasserstein space is separable and so is the space of optimal maps, by Lemma 2.4.6, so we may apply Lemma 2.4.8 and approximate $\mathbf{t}_{\theta_0}^{\Lambda}$ by simple maps for which the equality holds by linearity. On page 56 of the supplement, we show that these simple maps can be assumed optimal, and give the full details.

2.5 Bibliographical Notes

Our proof of Theorem 2.2.11 borrows heavily from Bolley et al. [29]. A similar result was obtained by Kloeckner [81], who also provides a lower bound of a similar order.

The origins of Sect. 2.3 can be traced back to the seminal work of Jordan et al. [74], who interpret the Fokker–Planck equation as a gradient flow (where functionals defined on $\mathscr{W}_2$ can be differentiated) with respect to the 2-Wasserstein metric. The Riemannian interpretation was (formally) introduced by Otto [99], and rigorously established by Ambrosio et al. [12] and others; see Villani [125, Chapter 15] for further bibliography and more details.

Compatible measures (Definition 2.3.1) were implicitly introduced by Boissard et al. [28] in the context of *admissible optimal maps* where one defines families of gradients of convex functions (T_i) such that $T_j^{-1} \circ T_i$ is a gradient of a convex function for any i and j. For (any) fixed measure $\gamma \in \mathscr{C}$, compatibility of $\mathscr{C}$ is then equivalent to admissibility of the collection of maps $\{\mathbf{t}_\gamma^\mu\}_{\mu \in \mathscr{C}}$. The examples we gave are also taken from [28].

Lemma 2.3.3 is from Cuesta-Albertos et al. [38, Theorem 2.9] (see also Zemel and Panaretos [135]).

Chapter 3
Fréchet Means in the Wasserstein Space $\mathscr{W}_2$

If H is a Hilbert space (or a closed convex subspace thereof) and $x_1, \dots, x_N \in H$, then the empirical mean $\overline{x}_N = N^{-1} \sum x_i$ is the unique element of H that minimises the sum of squared distances from the x_i's.[1] That is, if we define

$$F(\theta) = \sum_{i=1}^{N} \|\theta - x_i\|^2, \qquad \theta \in H,$$

then $\theta = \overline{x}_N$ is the unique minimiser of F. This is easily seen by "opening the squares" and writing

$$F(\theta) = F(\overline{x}_N) + N\|\theta - \overline{x}_N\|^2.$$

The concept of a Fréchet mean (Fréchet [55]) generalises the notion of mean to a more general metric space by replacing the usual "sum of squares" with a "sum of squared distances", giving rise to the so-called *Fréchet functional.* A closely related notion is that of a *Karcher mean* (Karcher [78]), a term that describes stationary points of the sum of squares functional, when the latter is differentiable (see Sect. 3.1.6). Population versions of Fréchet means, assuming the space is endowed with a probability law, can also be defined, replacing summation by expectation with respect to that law.

Electronic Supplementary Material The online version of this chapter (https://doi.org/10.1007/978-3-030-38438-8_3) contains supplementary material.

[1] It should be remarked that this is a Hilbertian property (or at least a property linked to an inner product), not merely a linear property. In other words, it does not extend to Banach spaces. As an example, let $H = \mathbb{R}^2$ with the L_1 norm and consider the vertices $(0,0)$, $(0,1)$, and $(1,0)$ of the unit simplex. The mean of these is $(1/3, 1/3)$ but for (x,y) in the triangle,

$$F(x,y) = (x+y)^2 + (x+1-y)^2 + (1-x+y)^2 = 2 + x^2 + y^2 + (x-y)^2$$

is minimised at $(0,0)$.

V. M. Panaretos, Y. Zemel, *An Invitation to Statistics in Wasserstein Space*, SpringerBriefs in Probability and Mathematical Statistics,
https://doi.org/10.1007/978-3-030-38438-8_3

Fréchet means are perhaps the most basic object of statistical interest, and this chapter studies such means when the underlying space is the Wasserstein space $\mathcal{W}_2$. In general, existence and uniqueness of a Fréchet mean can be subtle, but we will see that the nature of optimal transport allows for rather clean statements in the case of Wasserstein space.

3.1 Empirical Fréchet Means in $\mathcal{W}_2$

3.1.1 The Fréchet Functional

As foretold in the preceding paragraph, the definition of a Fréchet mean requires the definition of an appropriate sum-of-squares functional, the *Fréchet functional*:

Definition 3.1.1 (Empirical Fréchet Functional and Mean) *The Fréchet functional associated with measures $\mu^1,\dots,\mu^N \in \mathcal{W}_2(\mathcal{X})$ is*

$$F:\mathcal{W}_2(\mathcal{X})\to\mathbb{R} \qquad F(\gamma)=\frac{1}{2N}\sum_{i=1}^{N}W_2^2(\gamma,\mu^i), \qquad \gamma\in\mathcal{W}_2(\mathcal{X}). \tag{3.1}$$

A Fréchet mean of $(\mu^1,\dots,\mu^N)$ is a minimiser of F in $\mathcal{W}_2(\mathcal{X})$ (if it exists).

In analysis, a Fréchet mean is often called a *barycentre*. We shall use the terminology of "Fréchet mean" that is arguably more popular in statistics.[2]

The factor $1/(2N)$ is irrelevant for the definition of Fréchet mean. It is introduced in order to have simpler expressions for the derivatives (Theorems 3.1.14 and 3.2.13) and to be compatible with a population version $\mathbb{E}W_2^2(\gamma,\Lambda)/2$ (see (3.3)).

The first reference that deals with empirical Fréchet means in $\mathcal{W}_2(\mathbb{R}^d)$ is the seminal paper of Agueh and Carlier [2]. They treat the more general weighted Fréchet functional

$$F(\gamma)=\frac{1}{2}\sum_{i=1}^{N}w_iW_2^2(\gamma,\mu^i), \qquad 0\le w_i, \quad \sum_{i=1}^{N}w_i=1,$$

but, for simplicity, we shall focus on the case of equal weights. (If all the w_i's are rational, then the weighted functional can be encompassed in (3.1) by taking some of the μ^i's to be the same. The case of irrational w_i's is then treated with continuity arguments. Moreover, (3.3) encapsulates (3.1) as well as the weighted version when Λ can take finitely many values.)

[2] Interestingly, Fréchet himself [56] considered the Wasserstein metric between probability measures on $\mathbb{R}$, and some refer to this as the *Fréchet distance* (e.g., Dowson and Landau [44]), which is another reason to use this terminology.

3.1.2 Multimarginal Formulation, Existence, and Continuity

In [60], Gangbo and Święch consider the following *multimarginal* Monge–Kantorovich problem. Let $\mu^1,\dots,\mu^N$ be N measures in $\mathcal{W}_2(\mathcal{X})$ and let $\Pi(\mu^1,\dots,\mu^N)$ be the set of probability measures in $\mathcal{X}^N$ having $\{\mu^i\}_{i=1}^N$ as marginals. The problem is to minimise

$$G(\pi)=\frac{1}{2N^2}\int_{\mathcal{X}^N}\sum_{i<j}\|x_i-x_j\|^2\,\mathrm{d}\pi(x_1,\dots,x_N),\qquad \text{over}\quad \pi\in\Pi(\mu^1,\dots,\mu^N).$$

The factor $1/(2N^2)$ is of course irrelevant for the minimisation and its purpose will be clarified shortly. If $N=2$, we obtain the Kantorovich problem with quadratic cost. The probabilistic interpretation (as in Sect. 1.2) is that one is given random variables $X_1,\dots,X_N$ with marginal probability laws $\mu^1,\dots,\mu^N$ and seeks to construct a random vector $Y=(Y_1,\dots,Y_N)$ on $\mathcal{X}^N$ such that $X_i\stackrel{d}{=}Y_i$ and

$$\frac{1}{2N^2}\mathbb{E}\sum_{i<j}\|Y_i-Y_j\|^2\le\frac{1}{2N^2}\mathbb{E}\sum_{i<j}\|Z_i-Z_j\|^2.$$

for any other random vector $Z=(Z_1,\dots,Z_N)$ such that $X_i\stackrel{d}{=}Z_i$. Intuitively, we seek a random vector with prescribed marginals but maximally correlated entries.

We refer to elements of $\Pi(\mu^1,\dots,\mu^N)$ (equivalently, joint laws of $X_1,\dots,X_N$) as *multicouplings* (of $\mu^1,\dots,\mu^N$). Just like in the Kantorovich problem, there always exists an optimal multicoupling π.

Let us now show how the multimarginal problem is equivalent to the problem of finding the Fréchet mean of $\mu^1,\dots,\mu^N$. The first thing to observe is that the objective function can be written as

$$G(\pi)=\int_{\mathcal{X}^N}\frac{1}{2N}\sum_{i=1}^N\|x_i-M(x)\|^2\,\mathrm{d}\pi(x),\qquad M(x)=M(x_1,\dots,x_n)=\frac{1}{N}\sum_{i=1}^N x_i.$$

The next result shows that the Fréchet mean and the multicoupling problems are essentially the same.

Proposition 3.1.2 (Fréchet Means and Multicouplings) *Let $\mu^1,\dots,\mu^N\in\mathcal{W}(\mathcal{X})$. Then μ is a Fréchet mean of $(\mu^1,\dots,\mu^N)$ if and only if there exists an optimal multicoupling $\pi\in\mathcal{W}(\mathcal{X}^N)$ of $(\mu^1,\dots,\mu^N)$ such that $\mu=M\#\pi$, and furthermore $F(\mu)=G(\pi)$.*

Proof. Let π be an arbitrary multicoupling of $(\mu^1,\dots,\mu^N)$ and set $\mu=M\#\pi$. Then $(x\mapsto x_i,M)\#\pi$ is a coupling of μ^i and μ, and therefore

$$\int_{\mathcal{X}^N}\|x_i-M(x)\|^2\,\mathrm{d}\pi(x)\ge W^2(\mu,\mu_i).$$

Summation over i gives $F(\mu)\le G(\pi)$ and so $\inf F\le\inf G$.

For the other inequality, let $\mu \in \mathscr{W}(\mathscr{X})$ be arbitrary. For each i, let π^i be an optimal coupling between μ and μ^i. Invoking the gluing lemma (Ambrosio and Gigli [10, Lemma 2.1]), we may glue all π^i's using their common marginal μ. This procedure constructs a measure η on $\mathscr{X}^{N+1}$ with marginals $\mu_1,\dots,\mu_N,\mu$ and its relevant projection π is then a multicoupling of $\mu_1,\dots,\mu_N$.

Since $\mathscr{X}$ is a Hilbert space, the minimiser of $y \mapsto \sum \|x_i - y\|^2$ is $y = M(x)$. Thus

$$F(\mu)=\frac{1}{2N}\int_{\mathscr{X}^{N+1}}\sum_{i=1}^N \|x_i-y\|^2\,\mathrm{d}\eta(x,y)\ge\frac{1}{2N}\int_{\mathscr{X}^{N+1}}\sum_{i=1}^N \|x_i-M(x)\|^2\,\mathrm{d}\eta(x,y)=G(\pi).$$

In particular, $\inf F \ge \inf G$ and combining this with the established converse inequality we see that $\inf F = \inf G$. Observe also that the last displayed inequality holds as equality if and only if $y = M(x)$ η-almost surely, in which case $\mu = M\#\pi$. Therefore, if μ does not equal $M\#\pi$, then $F(\mu) > G(\pi) \ge F(M\#\pi)$, and μ cannot be optimal. Finally, if π is optimal, then

$$F(M\#\pi) \le G(\pi) = \inf G = \inf F$$

establishing optimality of $\mu = M\#\pi$ and completing the proof.

Since optimal couplings exist, we deduce that so do Fréchet means.

Corollary 3.1.3 (Fréchet Means and Moments) *Any finite collection of measures $\mu^1,\dots,\mu^N \in \mathscr{W}_2(\mathscr{X})$ admits a Fréchet mean μ, for all $p \ge 1$*

$$\int_{\mathscr{X}} \|x\|^p\,\mathrm{d}\mu(x) \le \frac{1}{N}\sum_{i=1}^N \int_{\mathscr{X}} \|x\|^p\,\mathrm{d}\mu^i(x),$$

and when $p > 1$ equality holds if and only if $\mu^1 = \cdots = \mu^N$.

Proof. Let π be a multicoupling of $\mu^1,\dots,\mu^N$ such that $\mu = M_N\#\pi$ (Proposition 3.1.2). Then

$$\int_{\mathscr{X}} \|x\|^p\,\mathrm{d}\mu(x) = \int_{\mathscr{X}^N} \left\|\frac{1}{N}\sum_{i=1}^N x_i\right\|^p \mathrm{d}\pi(x) \le \frac{1}{N}\sum_{i=1}^N \int_{\mathscr{X}^N} \|x_i\|^p\,\mathrm{d}\pi(x)$$
$$= \frac{1}{N}\sum_{i=1}^N \int_{\mathscr{X}} \|x\|^p\,\mathrm{d}\mu^i(x).$$

The statement about equality follows from strict convexity of $x \mapsto \|x\|^p$ if $p > 1$.

A further corollary of Proposition 3.1.2 is a bound on the support:

Corollary 3.1.4 *The support of any Fréchet mean is included in the set*

$$\frac{\operatorname{supp}\mu^1 + \cdots + \operatorname{supp}\mu^N}{N} = \left\{\frac{x_1 + \cdots + x_N}{N} : x_i \in \operatorname{supp}\mu^i\right\} \subseteq \operatorname{conv}\left(\bigcup_{i=1}^N \operatorname{supp}\mu^i\right).$$

In particular, if all the μ^i's are supported on a common convex set K, then so is any of their Fréchet means.

The multimarginal formulation also yields a continuity property for the empirical Fréchet mean. Conditions for uniqueness will be given in the next subsection.

Theorem 3.1.5 (Continuity of Fréchet Means) *Suppose that $W_2(\mu_k^i,\mu^i)\to 0$ for $i=1,\dots,N$ and let $\overline{\mu}_k$ denote any Fréchet mean of $(\mu_k^1,\dots,\mu_k^N)$. Then $(\overline{\mu}_k)$ stays in a compact set of $\mathscr{W}_2(\mathscr{X})$, and any limit point is a Fréchet mean of $(\mu^1,\dots,\mu^N)$.*

In particular, if $\mu^1,\dots,\mu^N$ have a *unique* Fréchet mean $\overline{\mu}$, then $\overline{\mu}_k\to\overline{\mu}$ in $\mathscr{W}_2(\mathscr{X})$.

Proof. We sketch the steps of the proof here, with the full details given on page 63 of the supplement.

Step 1: tightness of $(\overline{\mu}_k)$. This is true because the collection of multicouplings is tight, and the mean function M is continuous.

Step 2: weak limits are limits in $\mathscr{W}_2(\mathscr{X})$. This holds because the mean function has linear growth.

Step 3: the limit is a Fréchet mean of $(\mu^1,\dots,\mu^N)$. From Corollary 3.1.3, it follows that $\overline{\mu}_k$ must be sought on some fixed bounded set in $\mathscr{W}_2(\mathscr{X})$. On such sets, the Fréchet functionals are uniformly Lipschitz, so their minimisers converge as well.

3.1.3 Uniqueness and Regularity

A general situation in which Fréchet means are unique is when the Fréchet functional is strictly convex. In the Wasserstein space, this requires some regularity, but weak convexity holds in general. Absolutely continuous measures on infinite-dimensional $\mathscr{X}$ are defined in Definition 1.6.4.

Proposition 3.1.6 (Convexity of the Fréchet Functional) *Let $\Lambda,\gamma_i\in\mathscr{W}_2(\mathscr{X})$ and $t\in[0,1]$. Then*

$$W_2^2(t\gamma_1+(1-t)\gamma_2,\Lambda)\le tW_2^2(\gamma_1,\Lambda)+(1-t)W_2^2(\gamma_2,\Lambda). \tag{3.2}$$

When Λ is absolutely continuous, the inequality is strict unless $t\in\{0,1\}$ or $\gamma_1=\gamma_2$.

Remark 3.1.7 *The Wasserstein distance is not convex along geodesics. That is, if we replace the linear interpolant $t\gamma_1+(1-t)\gamma_2$ by McCann's interpolant, then $t\mapsto W_2^2(\gamma_t,\Lambda)$ is not necessarily convex (Ambrosio et al. [12, Example 9.1.5]).*

Proof. Let $\pi_i\in\Pi(\gamma_i,\Lambda)$ be optimal and notice that the linear interpolant $t\pi_1+(1-t)\pi_2\in\Pi(t\gamma_1+(1-t)\gamma_2,\Lambda)$, so that

$$W_2^2(t\gamma_1+(1-t)\gamma_2,\Lambda)\le\int_{\mathscr{X}^2}\|x-y\|^2\,\mathrm{d}[t\pi_1+(1-t)\pi_2](x,y),$$

which is (3.2). When Λ is absolutely continuous and $t \in (0,1)$, equality in (3.2) holds if and only if $\pi_t = t\pi_1 + (1-t)\pi_2 = (\mathbf{t}_\Lambda^{t\gamma_1+(1-t)\gamma_2} \times \mathbf{i})\#\Lambda$. But π_t is supported on the graphs of two functions: $\mathbf{t}_\Lambda^{\gamma_1}$ and $\mathbf{t}_\Lambda^{\gamma_2}$. Consequently, equality can hold only if these two maps equal Λ-almost surely, or, equivalently, if $\gamma_1 = \gamma_2$.

As a corollary, we deduce that the Fréchet mean is unique if one of the measures μ^i is absolutely continuous, and this extends to the population version (see Proposition 3.2.7).

We conclude this subsection by stating an important regularity property in the Euclidean case. See Agueh and Carlier [2, Proposition 5.1] for a proof.

Proposition 3.1.8 (L_∞-Regularity of Fréchet Means) *Let $\mu^1,\dots,\mu^N \in \mathcal{W}_2(\mathbb{R}^d)$ and suppose that μ^1 is absolutely continuous with density bounded by M. Then the Fréchet mean of $\{\mu^i\}$ is absolutely continuous with density bounded by N^dM and is consequently a Karcher mean.*

In Theorem 5.5.2, we extend Proposition 3.1.8 to the population level.

3.1.4 The One-Dimensional and the Compatible Case

When $\mathcal{X} = \mathbb{R}$, there is a simple expression for the Fréchet mean because $\mathcal{W}_2(\mathbb{R})$ can be imbedded in a Hilbert space. Indeed, recall that

$$W_2(\mu,\nu) = \|F_\mu^{-1} - F_\nu^{-1}\|_{L_2(0,1)}$$

(see Sect. 2.3.2 or 1.5). In view of that, $\mathcal{W}_2(\mathbb{R})$ can be seen as the convex closed subset of $L_2(0,1)$ formed by equivalence classes of left-continuous nondecreasing functions on $(0,1)$: any quantile function is left-continuous and nondecreasing, and any such function G can be seen to be the inverse function of the distribution function, the *right-continuous inverse* of G

$$F(x) = \inf\{t \in (0,1) : G(t) > x\} = \sup\{t \in (0,1) : G(t) \le x\}.$$

(See, for example, Bobkov and Ledoux [25, Appendix A].) Therefore, the Fréchet mean of $\mu^1,\dots,\mu^N \in \mathcal{W}_2(\mathbb{R})$ is the measure μ having quantile function

$$F_\mu^{-1} = \frac{1}{N}\sum_{i=1}^N F_{\mu^i}.$$

The Fréchet mean is thus unique. This is no longer true in higher dimension, unless some regularity is imposed on the measures (Proposition 3.2.7).

Boissard et al. [28] noticed that compatibility of $\mu^1,\dots,\mu^N$ according to Definition 2.3.1 allows for a simple solution to the Fréchet mean problem, as in the one-dimensional case. Recall from Proposition 3.1.2 that this is equivalent to the multimarginal problem. Returning to the original form of G, we obtain an easy lower bound for any $\pi \in \Pi(\mu^1,\dots,\mu^N)$:

$$G(\pi) = \frac{1}{2N^2}\int_{\mathcal{X}^N}\sum_{i<j}\|x_i - x_j\|^2\,\mathrm{d}\pi(x_1,\dots,x_N) \ge \frac{1}{2N^2}\sum_{i<j}W_2^2(\mu^i,\mu^j),$$

because the (i,j)-th marginal of π is a coupling of μ^i and μ^j. Thus, if equality above holds for π, then π is optimal and $M\#\pi$ is the Fréchet mean by Proposition 3.1.2. This is indeed the case for $\pi = (\mathbf{i}, \mathbf{t}_{\mu^1}^{\mu^2},\dots,\mathbf{t}_{\mu^1}^{\mu^N})\#\mu^1$ because the compatibility gives:

$$\begin{aligned}\int_{\mathcal{X}^N}\|x_i - x_j\|^2\,\mathrm{d}\pi(x_1,\dots,x_N) &= \int_{\mathcal{X}}\left\|\mathbf{t}_{\mu^1}^{\mu^i} - \mathbf{t}_{\mu^1}^{\mu^j}\right\|^2 \mathrm{d}\mu^1\\ &= \int_{\mathcal{X}}\left\|\mathbf{t}_{\mu^1}^{\mu^i}\circ\mathbf{t}_{\mu^j}^{\mu^i} - \mathbf{i}\right\| \mathrm{d}\mu^j = W_2^2(\mu^i,\mu^j).\end{aligned}$$

We may thus conclude, in a slightly more general form (γ was μ^1 above):

Theorem 3.1.9 (Fréchet Mean of Compatible Measures) *Suppose that* $\{\gamma,\mu^1,\dots,\mu^N\}$ *are compatible measures. Then*

$$\left[\frac{1}{N}\sum_{i=1}^{N}\mathbf{t}_{\gamma}^{\mu^i}\right]\#\gamma$$

is the Fréchet mean of $(\mu^1,\dots,\mu^N)$.

A population version is given in Theorem 5.5.3.

3.1.5 The Agueh–Carlier Characterisation

Agueh and Carlier [2] provide a useful sufficient condition for γ to be the Fréchet mean. When $\mathcal{X} = \mathbb{R}^d$, this condition is also necessary [2, Proposition 3.8], hence characterising Fréchet means in $\mathbb{R}^d$. It will allow us to easily deduce some equivariance results for Fréchet means with respect to independence (Lemma 3.1.11) and rotations (3.1.12). More importantly, it provides a sufficient condition under which a local minimum of F is a global minimum (Theorem 3.1.15) and the same idea can be used to relate the population Fréchet mean to the expected value of the optimal maps (Theorem 4.2.4). Recall that ϕ^* denotes the Legendre transform of ϕ, as defined on page 14.

Proposition 3.1.10 (Fréchet Means and Potentials) *Let* $\mu^1,\dots,\mu^N \in \mathcal{W}_2(\mathcal{X})$ *be absolutely continuous, let* $\gamma \in \mathcal{W}_2(\mathcal{X})$ *and denote by* ϕ_i^* *the convex potentials of* $\mathbf{t}_{\mu^i}^{\gamma}$. *If* $\phi_i = \phi_i^{**}$ *are such that*

$$\frac{1}{N}\sum_{i=1}^{N}\phi_i(x) \le \frac{1}{2}\|x\|^2, \qquad \forall x\in\mathcal{X}, \qquad \textit{with equality } \gamma\textit{-almost surely},$$

then γ *is the unique Fréchet mean of* $\mu^1,\dots,\mu^N$.

Proof. Uniqueness follows from Proposition 3.2.7. If $\theta \in \mathscr{W}_2(\mathscr{X})$ is any measure, then the Kantorovich duality yields

$$W_2^2(\gamma,\mu^i) = \int_{\mathscr{X}} \left(\frac{1}{2}\|x\|^2 - \phi_i(x)\right) \mathrm{d}\gamma(x) + \int_{\mathscr{X}} \left(\frac{1}{2}\|y\|^2 - \phi_i^*(y)\right) \mathrm{d}\mu^i(y);$$

$$W_2^2(\theta,\mu^i) \geq \int_{\mathscr{X}} \left(\frac{1}{2}\|x\|^2 - \phi_i(x)\right) \mathrm{d}\theta(x) + \int_{\mathscr{X}} \left(\frac{1}{2}\|y\|^2 - \phi_i^*(y)\right) \mathrm{d}\mu^i(y).$$

Summation over i gives the result.

A population version of this result, based on similar calculations, is given in Theorem 4.2.4.

The next two results are formulated in $\mathbb{R}^d$ because then the converse of Proposition 3.1.10 is proven to be true. If one could extend [2, Proposition 3.8] to any separable Hilbert $\mathscr{X}$, then the two lemmata below will hold with $\mathbb{R}^d$ replaced by $\mathscr{X}$. The simple proofs are given on page 66 of the supplement.

Lemma 3.1.11 (Independent Fréchet Means) *Let $\mu^1,\dots,\mu^N$ and $\nu^1,\dots,\nu^N$ be absolutely continuous measures in $\mathscr{W}_2(\mathbb{R}^{d_1})$ and $\mathscr{W}_2(\mathbb{R}^{d_2})$ with Fréchet means μ and ν, respectively. Then the independent coupling $\mu\otimes\nu$ is the Fréchet mean of $\mu^1\otimes\nu^1,\dots,\mu^N\otimes\nu^N$.*

By induction (or a straightforward modification of the proof), one can show that the Fréchet mean of $(\mu^i\otimes\nu^i\otimes\rho^i)$ is $\mu\otimes\nu\otimes\rho$, and so on.

Lemma 3.1.12 (Rotated Fréchet Means) *If μ is the Fréchet mean of the absolutely continuous measures $\mu^1,\dots,\mu^N$ and U is orthogonal, then $U\#\mu$ is the Fréchet mean of $U\#\mu^1,\dots,U\#\mu^N$.*

3.1.6 Differentiability of the Fréchet Functional and Karcher Means

Since we seek to minimise the Fréchet functional F, it would be helpful if F were differentiable, because we could then find at least local minima by solving the equation $F'=0$. This observation of Karcher [78] leads to the notion of *Karcher mean.*

Definition 3.1.13 (Karcher Mean) *Let F be a Fréchet functional associated with some random measure Λ in $\mathscr{W}_2(\mathscr{X})$. Then γ is a Karcher mean for Λ if F is differentiable at γ and $F'(\gamma)=0$.*

Of course, if γ is a Fréchet mean for the random measure Λ and F is differentiable at γ, then $F'(\gamma)$ must vanish. In this subsection, we build upon the work of Ambrosio et al. [12] and determine the derivative of the Fréchet functional. This will not only allow for a simple characterisation of Karcher means in terms of the optimal maps $\mathbf{t}_\gamma^\Lambda$ (Proposition 3.2.14), but will also be the cornerstone of the construction of a steepest descent algorithm for empirical calculation of Fréchet means. The differentiability holds at the population level too (Theorem 3.2.13).

It turns out that the tangent bundle structure described in Sect. 2.3 gives rise to a differentiable structure in the Wasserstein space. Fix $\mu^0 \in \mathscr{W}_2(\mathscr{X})$ and consider the function

$$F_0 : \mathscr{W}_2(\mathscr{X}) \to \mathbb{R}, \qquad F_0(\gamma) = \frac{1}{2} W_2^2(\gamma, \mu^0).$$

Ambrosio et al. [12, Corollary 10.2.7] show that when γ is absolutely continuous,

$$\lim_{W_2(\nu,\gamma)\to 0} \frac{F_0(\nu) - F_0(\gamma) + \int_{\mathscr{X}} \left\langle \mathbf{t}_\gamma^{\mu^0}(x) - x, \mathbf{t}_\gamma^\nu(x) - x \right\rangle \mathrm{d}\gamma(x)}{W_2(\nu,\gamma)} = 0.$$

Parts of the proof of this result (the limit superior above is ≤ 0; the limit inferior is bounded below) are reproduced in Proposition 3.2.12. The integral above can be seen as the inner product

$$\left\langle \mathbf{t}_\gamma^{\mu^0} - \mathbf{i}, \mathbf{t}_\gamma^\nu - \mathbf{i} \right\rangle$$

in the space $\mathscr{L}_2(\gamma)$ that includes as a (closed) subspace the tangent space Tan_γ. In terms of this inner product and the log map, we can write

$$F_0(\nu) - F_0(\gamma) = -\left\langle \log_\gamma(\mu^0), \log_\gamma(\nu) \right\rangle + o(W_2(\nu,\gamma)), \qquad \nu \to \gamma \quad \text{in } \mathscr{W}_2,$$

so that F_0 is Fréchet-differentiable[3] at γ with derivative

$$F_0'(\gamma) = -\log_\gamma(\mu^0) = -\left(\mathbf{t}_\gamma^{\mu^0} - \mathbf{i}\right) \in \mathrm{Tan}_\gamma.$$

By linearity, one immediately obtains:

Theorem 3.1.14 (Gradient of the Fréchet Functional) *Fix a collection of measures $\mu^1, \dots, \mu^N \in \mathscr{W}_2(\mathscr{X})$. When $\gamma \in \mathscr{W}_2(\mathscr{X})$ is absolutely continuous, the Fréchet functional*

$$F(\gamma) = \frac{1}{2N} \sum_{i=1}^N W_2^2(\gamma, \mu^i), \qquad \gamma \in \mathscr{W}_2(\mathscr{X})$$

is Fréchet-differentiable and

$$F'(\gamma) = -\frac{1}{N} \sum_{i=1}^N \log_\gamma(\mu^i) = -\frac{1}{N} \sum_{i=1}^N \left(\mathbf{t}_\gamma^{\mu_i} - \mathbf{i}\right).$$

It follows from this that an absolutely continuous $\gamma \in \mathscr{W}_2(\mathscr{X})$ is a Karcher mean if and only if the average of the optimal maps is the identity. If in addition one μ^i is absolutely continuous with bounded density, then the Fréchet mean $\overline{\mu}$ is absolutely

[3] The notion of Fréchet derivative is also named after Maurice Fréchet, but is not directly related to Fréchet means.

continuous by Proposition 3.1.8, so it is a Karcher mean. The result extends to the population version; see Proposition 3.2.14.

It may happen that a collection $\mu^1,\dots,\mu^N$ of absolutely continuous measures have a Karcher mean that is not a Fréchet mean; see Álvarez-Esteban et al. [9, Example 3.1] for an example in $\mathbb{R}^2$. But a Karcher mean γ is "almost" a Fréchet mean in the following sense. By Proposition 3.2.14, $N^{-1}\sum \mathbf{t}_\gamma^{\mu^i}(x) = x$ for γ-almost all x. If, on the other hand, the equality holds *for all* $x \in \mathcal{X}$, then γ is the Fréchet mean by taking integrals and applying Proposition 3.1.10. One can hope that under regularity conditions, the γ-almost sure equality can be upgraded to equality everywhere. Indeed, this is the case:

Theorem 3.1.15 (Optimality Criterion for Karcher Means) *Let $U \subseteq \mathbb{R}^d$ be an open convex set and let $\mu^1,\dots,\mu^N \in \mathcal{W}_2(\mathbb{R}^d)$ be probability measures on U with bounded strictly positive densities $g^1,\dots,g^N$. Suppose that an absolutely continuous Karcher mean γ is supported on U with bounded strictly positive density f there. Then γ is the Fréchet mean of $\mu^1,\dots,\mu^N$ if one of the following holds:*

1. *$U = \mathbb{R}^d$ and the densities $f, g^1,\dots,g^N$ are of class $C^{0,\alpha}$ for some $\alpha > 0$;*
2. *U is bounded and the densities $f, g^1,\dots,g^N$ are bounded below on U.*

Proof. The result exploits Caffarelli's regularity theory for Monge–Ampère equations in the form of Theorem 1.6.7. In the first case, there exist C^1 (in fact, $C^{2,\alpha}$) convex potentials φ_i on $\mathbb{R}^d$ with $\mathbf{t}_\gamma^{\mu^i} = \nabla\varphi_i$, so that $\mathbf{t}_\gamma^{\mu^i}(x)$ is a singleton for all $x \in \mathbb{R}^d$. The set $\{x \in \mathbb{R}^d : \sum \mathbf{t}_\gamma^{\mu^i}(x)/N \neq x\}$ is γ-negligible (and hence Lebesgue negligible) and open by continuity. It is therefore empty, so $F'(\gamma) = 0$ everywhere, and γ is the Fréchet mean (see the discussion before the theorem).

In the second case, by the same argument we have $\sum \mathbf{t}_\gamma^{\mu^i}(x)/N = x$ for all $x \in U$. Since U is convex, there must exist a constant C such that $\sum \varphi_i(x) = C + N\|x\|^2/2$ for all $x \in U$, and we may assume without loss of generality that $C = 0$. If one repeats the proof of Proposition 3.1.10, then $F(\gamma) \le F(\theta)$ for all $\theta \in P(U)$. By continuity considerations, the inequality holds for all $\theta \in P(\overline{U})$ (Theorem 2.2.7) and since $\overline{U}$ is closed and convex, γ is the Fréchet mean by Corollary 3.1.3.

3.2 Population Fréchet Means

In this section, we extend the notion of empirical Fréchet mean to the population level, where Λ is a random element in $\mathcal{W}_2(\mathcal{X})$ (a measurable mapping from a probability space to $\mathcal{W}_2(\mathcal{X})$). This requires a different strategy, since it is not clear how to define the analogue of the multicouplings at that level of abstraction. However, it is important to point out that when there is more structure in Λ, multicouplings can be defined as laws of stochastic processes; see Pass [102] for a detailed account of the problem in this case.

In analogy with (3.1), we define:

Definition 3.2.1 (Population Fréchet Mean) *Let Λ be a random measure in $\mathscr{W}_2(\mathscr{X})$. The Fréchet mean of Λ is the minimiser (if it exists and is unique) of the Fréchet functional*

$$F(\gamma) = \frac{1}{2}\mathbb{E}W_2^2(\gamma,\Lambda), \qquad \gamma \in \mathscr{W}_2(\mathscr{X}). \tag{3.3}$$

Since W_2 is continuous and nonnegative, the expectation is well-defined.

3.2.1 Existence, Uniqueness, and Continuity

Existence and uniqueness of Fréchet means on a general metric space M are rather delicate questions. Usually, existence proofs are easier: for example, since the Fréchet functional F is continuous on M (as we show below), one often invokes local compactness of M in order to establish existence of a minimiser. Unfortunately, a different strategy is needed when $M = \mathscr{W}_2(\mathscr{X})$, because the Wasserstein space is not locally compact (Proposition 2.2.9).

The first thing to notice is that F is indeed continuous (this is clear for the empirical version). This is a consequence of the triangle inequality and holds when $\mathscr{W}_2(\mathscr{X})$ is replaced by any metric space.

Lemma 3.2.2 (Finiteness of the Fréchet Functional) *If F is not identically infinite, then it is finite and locally Lipschitz everywhere on $\mathscr{W}_2(\mathscr{X})$.*

Proof. Assume that F is finite at γ. If θ is any other measure in $\mathscr{W}_2(\mathscr{X})$, write

$$2F(\gamma) - 2F(\theta) = \mathbb{E}[W_2(\gamma,\Lambda) - W_2(\theta,\Lambda)][W_2(\gamma,\Lambda) + W_2(\theta,\Lambda)].$$

Since $x \le 1 + x^2$ for all x, the triangle inequality in $\mathscr{W}_2(\mathscr{X})$ yields

$$\begin{aligned}2|F(\gamma) - F(\theta)| &\le W_2(\gamma,\theta)[2\mathbb{E}W_2(\gamma,\Lambda) + W_2(\theta,\gamma)] \\ &\le W_2(\gamma,\theta)[2\mathbb{E}W_2^2(\gamma,\Lambda) + 2 + W_2(\theta,\gamma)].\end{aligned}$$

Since $F(\gamma) < \infty$, this shows that F is finite everywhere and the right-hand side vanishes as $\theta \to \gamma$ in $\mathscr{W}_2(\mathscr{X})$. Now that we know that F is continuous, the same upper bound shows that it is in fact locally Lipschitz.

Example: let (a_n) be a sequence of positive numbers that sum up to one. Let $x_n = 1/a_n$ and suppose that Λ equals $\delta\{x_n\} \in \mathscr{W}_2(\mathbb{R})$ with probability a_n. Then

$$\mathbb{E}W_2^2(\Lambda,\delta_0) = \sum_{n=1}^{\infty} a_n x_n^2 = \sum_{n=1}^{\infty} 1/a_n = \infty,$$

and by Lemma 3.2.2 F is identically infinite. Henceforth, we say that F is finite when the condition in Lemma 3.2.2 holds.

Using the lower semicontinuity (2.5), one can prove existence on $\mathbb{R}^d$ rather easily. (The empirical means exist even in infinite dimensions by Corollary 3.1.3.)

Proposition 3.2.3 (Existence of Fréchet Means) *The Fréchet functional associated with any random measure Λ in $\mathscr{W}_2(\mathbb{R}^d)$ admits a minimiser.*

Proof. The assertion is clear if F is identically infinite. Otherwise, let (γ_n) be a minimising sequence. We wish to show that the sequence is tight. Define $L = \sup_n F(\gamma_n) < \infty$ and observe that since $x \le 1 + x^2$ for all $x \in \mathbb{R}$,

$$\mathbb{E}W_2(\gamma_n, \Lambda) \le 1 + \mathbb{E}W_2^2(\gamma_n, \Lambda) \le 2L + 1, \qquad n = 1, 2, \dots.$$

By the triangle inequality

$$L' = \mathbb{E}W_2(\delta_0, \Lambda) \le W_2(\delta_0, \gamma_1) + \mathbb{E}W_2(\gamma_1, \Lambda) \le W_2(\delta_0, \gamma_1) + 2L + 1$$

so that for all n

$$\left(\int_{\mathbb{R}^d} \|x\|^2 \, d\gamma_n(x) \right)^{1/2} = W_2(\gamma_n, \delta_0) \le \mathbb{E}W_2(\gamma_n, \Lambda) + \mathbb{E}W_2(\Lambda, \delta_0) \le 2L + 1 + L' < \infty.$$

Since closed and bounded sets in $\mathbb{R}^d$ are compact, it follows that (γ_n) is a tight sequence. We may assume that $\gamma_n \to \gamma$ weakly, then use (2.5) and Fatou's lemma to obtain

$$2F(\gamma) = \mathbb{E}W_2^2(\gamma, \Lambda) \le \mathbb{E} \liminf_{n\to\infty} W_2^2(\gamma_n, \Lambda) \le \liminf_{n\to\infty} \mathbb{E}W_2^2(\gamma_n, \Lambda) = 2 \inf F.$$

Thus, γ is a minimiser of F, and existence is established.

When $\mathscr{X}$ is an infinite-dimensional Hilbert space, existence still holds under a compactness assumption. We first prove a result about the support of the Fréchet mean. At the empirical level, one can say more about the support (see Corollary 3.1.4).

Proposition 3.2.4 (Support of Fréchet Mean) *Let Λ be a random measure in $\mathscr{W}_2(\mathscr{X})$ and let $K \subseteq \mathscr{X}$ be a convex closed set such that $\mathbb{P}[\Lambda(K) = 1] = 1$. If γ minimises F, then $\gamma(K) = 1$.*

Remark 3.2.5 *For any closed $K \subseteq \mathscr{X}$ and any $\alpha \in [0,1]$, the set $\{\Lambda \in \mathscr{W}_p(\mathscr{X}) : \Lambda(K) \ge \alpha\}$ is closed in $\mathscr{W}_p(\mathscr{X})$ because $\{\Lambda \in P(\mathscr{X}) : \Lambda(K) \ge \alpha\}$ is weakly closed by the portmanteau lemma (Lemma 1.7.1).*

The proof amounts to a simple projection argument; see page 70 in the supplement.

Corollary 3.2.6 *If there exists a compact convex K satisfying the hypothesis of Proposition 3.2.4, then the Fréchet functional admits a minimiser supported on K.*

Proof. Proposition 3.2.4 allows us to restrict the domain of F to $\mathscr{W}_2(K)$, the collection of probability measures supported on K. Since this set is compact in $\mathscr{W}_2(\mathscr{X})$ (Corollary 2.2.5), the result follows from continuity of F.

From the convexity (3.2), one obtains a simple criterion for uniqueness. See Definition 1.6.4 for absolute continuity in infinite dimensions.

Proposition 3.2.7 (Uniqueness of Fréchet Means) *Let Λ be a random measure in $\mathscr{W}_2(\mathscr{X})$ with finite Fréchet functional. If Λ is absolutely continuous with positive (inner) probability, then the Fréchet mean of Λ is unique (if it exists).*

Remark 3.2.8 *It is not obvious that the set of absolutely continuous measures is measurable in $\mathscr{W}_2(\mathscr{X})$. We assume that there exists a Borel set $A \subset \mathscr{W}_2(\mathscr{X})$ such that $\mathbb{P}(\Lambda \in A) > 0$ and all measures in A are absolutely continuous.*

Proof. By taking expectations in (3.2), one sees that F is convex on $\mathscr{W}_2(\mathscr{X})$ with respect to linear interpolants. From Proposition 3.1.6, we conclude that

$$\Lambda \text{ absolutely continuous} \quad \Longrightarrow \quad \gamma \mapsto \frac{1}{2}W_2^2(\gamma,\Lambda) \text{ strictly convex.}$$

As F was already shown to be weakly convex in any case, it follows that

$$\mathbb{P}(\Lambda \text{ absolutely continuous}) > 0 \quad \Longrightarrow \quad F \text{ strictly convex.}$$

Since strictly convex functionals have at most one minimiser, this completes the proof.

We state without proof an important consistency result (Le Gouic and Loubes [87, Theorem 3]). Since $\mathscr{W}_2(\mathscr{X})$ is a complete and separable metric space, we can define the "second degree" Wasserstein space $\mathscr{W}_2(\mathscr{W}_2(\mathscr{X}))$. The law of a random measure Λ is in $\mathscr{W}_2(\mathscr{W}_2(\mathscr{X}))$ if and only if the corresponding Fréchet functional is finite.

Theorem 3.2.9 (Consistency of Fréchet Means) *Let Λ_n, Λ be random measures in $\mathscr{W}_2(\mathbb{R}^d)$ with finite Fréchet functionals and laws $\mathbb{P}_n, \mathbb{P} \in \mathscr{W}_2(\mathscr{W}_2(\mathbb{R}^d))$. If $\mathbb{P}_n \to \mathbb{P}$ in $\mathscr{W}_2(\mathscr{W}_2(\mathbb{R}^d))$, then any sequence λ_n of Fréchet means of Λ_n has a W_2-limit point λ, which is a Fréchet mean of Λ.*

See the Bibliographical Notes for a more general formulation.

Corollary 3.2.10 (Wasserstein Law of Large Numbers) *Let Λ be a random measure in $\mathscr{W}_2(\mathbb{R}^d)$ with finite Fréchet functional and let $\Lambda_1, \ldots$ be a sample from Λ. Assume λ is the unique Fréchet mean of Λ (see Proposition 3.2.7). Then almost surely, the sequence of empirical Fréchet means of $\Lambda_1, \ldots, \Lambda_n$ converges to λ.*

Proof. Let $\mathbb{P}$ be the law of Λ and let $\mathbb{P}_n$ be its empirical counterpart (a random element in $\mathscr{W}_2(\mathscr{W}_2(\mathbb{R}^d))$. Like in the proof of Proposition 2.2.6 (with $\mathscr{X}$ replaced by the complete separable metric space $\mathscr{W}_2(\mathbb{R}^d)$), almost surely $\mathbb{P}_n \to \mathbb{P}$ in $\mathscr{W}_2(\mathscr{W}_2(\mathbb{R}^d))$ and Theorem 3.2.9 applies.

Under a compactness assumption, one can give a direct proof for the law of large numbers as in Theorem 3.1.5. This is done on page 71 in the supplement.

3.2.2 The One-Dimensional Case

As a generalisation of the empirical version, we have:

Theorem 3.2.11 (Fréchet Means in $\mathcal{W}_2(\mathbb{R})$) *Let Λ be a random measure in $\mathcal{W}_2(\mathbb{R})$ with finite Fréchet functional. Then the Fréchet mean of Λ is the unique measure λ with quantile function $F_\lambda^{-1}(t) = \mathbb{E}F_\Lambda^{-1}(t)$, $t \in (0,1)$.*

Proof. Since $L_2(0,1)$ is a Hilbert space, the random element $F_\Lambda^{-1} \in L_2(0,1)$ has a unique Fréchet mean $g \in L_2(0,1)$, defined by the relations $\langle g, f\rangle = \mathbb{E}\left\langle F_\Lambda^{-1}, f\right\rangle$ for all $f \in L_2(0,1)$. On page 72 of the supplement, we show that g can be identified with F_λ^{-1}.

Interestingly, no regularity is needed in order for the Fréchet mean to be unique. This is not the case for higher dimensions, see Proposition 3.2.7. If there is some regularity, then one can state Theorem 3.2.11 in terms of optimal maps, because F_Λ^{-1} is the optimal map from $\mathrm{Leb}|_{[0,1]}$ to Λ. If $\gamma \in \mathcal{W}_2(\mathbb{R})$ is any absolutely continuous (or even just continuous) measure, then Theorem 3.2.11 can be stated as follows: the Fréchet mean of Λ is the measure $[\mathbb{E}\mathbf{t}_\gamma^\Lambda]\#\gamma$. A generalisation of this result to compatible measures (Definition 2.3.1) can be carried out in the same way, since compatible measures are imbedded in a Hilbert space, using the Bochner integrals for the definition of the expected optimal maps (see Sect. 2.4).

3.2.3 Differentiability of the Population Fréchet Functional

We now use the Fubini result (Proposition 2.4.9) in order to extend the differentiability of the Fréchet functional to the population version. This will follow immediately if we can interchange the expectation and the derivative in the form

$$F'(\gamma) = \frac{1}{2}(\mathbb{E}W_2^2)'(\gamma,\Lambda) = \mathbb{E}\left(\frac{1}{2}W_2^2\right)'(\gamma,\Lambda) = -\mathbb{E}(\mathbf{t}_\gamma^\Lambda - \mathbf{i}).$$

In order to do this, we will use dominated convergence in conjunction with uniform bounds on the slopes

$$u(\theta,\Lambda) = \frac{0.5W_2^2(\theta,\Lambda) - 0.5W_2^2(\theta_0,\Lambda) + \int_{\mathcal{X}}\langle \mathbf{t}_{\theta_0}^\Lambda - i, \mathbf{t}_{\theta_0}^\theta - i\rangle\,\mathrm{d}\theta_0}{W_2(\theta,\theta_0)}, \qquad u(\theta_0,\Lambda) = 0. \tag{3.4}$$

Proposition 3.2.12 (Slope Bounds) *Let θ_0, Λ, and θ be probability measures with θ_0 absolutely continuous, and set $\delta = W_2(\theta,\theta_0)$. Then*

$$\frac{1}{2}\delta - W_2(\theta_0,\Lambda) - \sqrt{2W_2^2(\theta_0,\delta_0) + 2W_2^2(\Lambda,\delta_0)} \le u(\theta,\Lambda) \le \frac{1}{2}\delta,$$

where u is defined by (3.4). *If the measures are compatible in the sense of Definition 2.3.1, then $u(\theta,\Lambda)=\delta/2$.*

The proof is a slight variation of Ambrosio et al. [12, Theorem 10.2.2 and Proposition 10.2.6], and the details are given on page 72 of the supplement.

Theorem 3.2.13 (Population Fréchet Gradient) *Let Λ be a random measure with finite Fréchet functional F. Then F is Fréchet-differentiable at any absolutely continuous θ_0 in the Wasserstein space, and $F'(\theta_0)=\mathbb{E}\mathbf{t}_{\theta_0}^{\Lambda}-\mathbf{i}\in\mathscr{L}_2(\theta_0)$. More precisely,*

$$\frac{F(\theta)-F(\theta_0)+\int_{\mathscr{X}}\langle\mathbb{E}\mathbf{t}_{\theta_0}^{\Lambda}-\mathbf{i},\mathbf{t}_{\theta_0}^{\theta}-\mathbf{i}\rangle\,\mathrm{d}\theta_0}{W_2(\theta,\theta_0)}\to 0,\qquad \theta\to\theta_0\quad \textit{in } \mathscr{W}_2.$$

Thus, the Fréchet derivative of F can be identified with the map $-(\mathbb{E}\mathbf{t}_{\theta_0}^{\Lambda}-\mathbf{i})$ in the tangent space at θ_0, a subspace of $\mathscr{L}_2(\theta_0)$.

Proof. Introduce the slopes $u(\theta,\Lambda)$ defined by (3.4). Then for all $\Lambda, u(\theta,\Lambda)\to 0$ as $W_2(\theta,\theta_0)\to 0$, by the differentiability properties established above. Let us show that $\mathbb{E}u(\theta,\Lambda)\to 0$ as well. By Proposition 3.2.12, the expectation of u is bounded above by a constant that does not depend on Λ, and below by the negative of

$$\begin{aligned}&\mathbb{E}W_2(\theta_0,\Lambda)+\mathbb{E}\sqrt{2W_2^2(\theta_0,\delta_0)+2W_2^2(\Lambda,\delta_0)}\\&\quad\le\sqrt{2}W_2(\theta_0,\delta_0)+\mathbb{E}W_2(\theta_0,\Lambda)+\sqrt{2}\mathbb{E}W_2(\Lambda,\delta_0).\end{aligned}$$

Both expectations are finite by the hypothesis on Λ because the Fréchet functional is finite. The dominated convergence theorem yields

$$\mathbb{E}u(\theta,\Lambda)=\frac{F(\theta)-F(\theta_0)+\mathbb{E}\int_{\mathscr{X}}\langle\mathbf{t}_{\theta_0}^{\Lambda}-i,\mathbf{t}_{\theta_0}^{\theta}-i\rangle\,\mathrm{d}\theta_0}{W_2(\theta,\theta_0)}\to 0,\qquad W_2(\theta_0,\theta)\to 0.$$

The measurability of the integral and the result then follow from Fubini's theorem (see Proposition 2.4.9).

Proposition 3.2.14 *Let Λ be a random measure in $\mathscr{W}_2(\mathscr{X})$ with finite Fréchet functional F, and let γ be absolutely continuous in $\mathscr{W}_2(\mathscr{X})$. Then γ is a Karcher mean of Λ if and only if $\mathbb{E}\mathbf{t}_{\gamma}^{\Lambda}-\mathbf{i}=0$ in $\mathscr{L}_2(\gamma)$. Furthermore, if γ is a Fréchet mean of Λ, then it is also a Karcher mean.*

The characterisation of Karcher means follows immediately from Theorem 3.2.13. The other statement is that the derivative vanishes at the minimum, which is fairly obvious intuitively; see page 73 in the supplement.

3.3 Bibliographical Notes

Proposition 3.1.2 is essentially due to Agueh and Carlier [2, Proposition 4.2], who show it on $\mathbb{R}^d$ (see also Zemel and Panaretos [134, Theorem 2]). An earlier result in a compact setting can be found in Carlier and Ekeland [33]. The formulation given here is from Masarotto et al. [91]. A more general version is provided by Le Gouic and Loubes [87, Theorem 8].

Lemmata 3.1.11 and 3.1.12 are from [135], but were known earlier (e.g., Bonneel et al. [30]).

Proposition 3.1.6 is a simplified version of Álvarez-Esteban et al. [8, Theorem 2.8] (see [8, Corollary 2.9]).

Propositions 3.2.3 and 3.2.7 are from Bigot and Klein [22], who also show the law of large numbers (Corollary 3.2.10) and deal with the one-dimensional setup (Theorem 3.2.11) in a compact setting. Section 2.4 appears to be new, but see the discussion in its beginning for other measurability results.

Barycentres can be defined for any $p \geq 1$ as the measures minimising $\mu \mapsto \mathbb{E}W_p^p(\Lambda, \mu)$. (Strictly speaking, these are not Fréchet means unless $p = 2$.) Le Gouic and Loubes [87] show Proposition 3.2.3 and Theorem 3.2.9 in this more general setup, where $\mathbb{R}^d$ can be replaced by any separable locally compact geodesic space.

Chapter 4
Phase Variation and Fréchet Means

Why is it relevant to construct the Fréchet mean of a collection of measures with respect to the Wasserstein metric? A simple answer is that this kind of average will often express a more natural notion of "typical" realisation of a random probability distribution than an arithmetic average.[1] Much more can be said, however, in that the Wasserstein–Fréchet mean and the closely related notion of an optimal multicoupling arise canonically as the appropriate framework for the formulation and solution to the problem of *separation of amplitude and phase variation of a point process*. It would almost seem that Wasserstein–Fréchet means were "made" for precisely this problem.

When analysing the (co)variation of a real-valued stochastic process $\{Y(x) : x \in K\}$ over a convex compact domain K, it can be broadly said that one may distinguish two layers of variation:

- *Amplitude variation*. This is the "classical" variation that one would also encounter in multivariate analysis, and refers to the stochastic fluctuations around a mean level, usually encoded in the covariance kernel, at least up to second order.

  In short, this is variation "in the y-axis" (ordinate).

Electronic Supplementary Material The online version of this chapter (https://doi.org/10.1007/978-3-030-38438-8_4) contains supplementary material.

---

[1] For instance, the arithmetic average of two scalar Gaussians $N(\mu_1, 1)$ and $N(\mu_2, 1)$ will be their mixture with equal weights, but their Fréchet–Wasserstein average will be the Gaussian $N(\frac{1}{2}\mu_1 + \frac{1}{2}\mu_2, 1)$ (see Lemma 4.2.1), which is arguably more representative from an intuitive point of view. In much the same way, the Fréchet–Wasserstein average of probability measures representing some type of object (e.g., normalised greyscale images of faces) will also be an object of the same type. This sort of phenomenon is well-known in manifold statistics, more generally, and is arguably one of the key motivations to account for the non-linear geometry of the sample space, rather than imbed it into a larger linear space and use the addition operation.

V. M. Panaretos, Y. Zemel, *An Invitation to Statistics in Wasserstein Space*, SpringerBriefs in Probability and Mathematical Statistics,
https://doi.org/10.1007/978-3-030-38438-8_4

- *Phase variation*. This is a second layer of non-linear variation peculiar to continuous domain stochastic processes, and is rarely—if ever—encountered in multivariate analysis. It arises as the result of random changes (or deformations) in the time scale (or the spatial domain) of definition of the process. It can be conceptualised as a composition of the stochastic process with a random transformation (warp map) acting on its domain.

  This is variation "in the x-axis" (abscissa).

The terminology on amplitude/phase variation is adapted from random trigonometric functions, which may vary in amplitude (oscillations in the range of the function) or phase (oscillations in the domain of the function). Failing to properly account for the superposition of these two forms of variation may entirely distort the findings of a statistical analysis of the random function (see Sect. 4.1.1). Consequently, it is an important problem to be able to separate the two, thus correctly accounting for the distinct contribution of each. The problem of separation is also known as that of *registration* (Ramsay and Li [108]), *synchronisation* (Wang and Gasser [129]), or *multireference alignment* (Bandeira et al. [16]), though in some cases these terms refer to a simpler problem where there is no amplitude variation at all.

Phase variation naturally arises in the study of random phenomena where there is no absolute notion of time or space, but every realisation of the phenomenon evolves according to a time scale that is intrinsic to the phenomenon itself, and (unfortunately) unobservable. Processes related to physiological measurements, such as *growth curves* and *neuronal signals*, are usual suspects. Growth curves can be modelled as *continuous random functions (functional data)*, whereas neuronal signals are better modelled as *discrete random measures (point processes)*. We first describe amplitude/phase variation in the former[2] case, as that is easier to appreciate, before moving on to the latter case, which is the main subject of this chapter.

4.1 Amplitude and Phase Variation

4.1.1 The Functional Case

Let K denote the unit cube $[0,1]^d \subset \mathbb{R}^d$. A real random function $Y = (Y(x) : x \in K)$ can, broadly speaking, have two types of variation. The first, *amplitude variation*, results from $Y(x)$ being a random variable for every x and describes its fluctuations around the mean level $m(x) = \mathbb{E}Y(x)$, usually encoded by the variance $\mathrm{var}Y(x)$. For this reason, it can be referred to as "variation in the y-axis". More

[2] As the functional case will only serve as a motivation, our treatment of this case will mostly be heuristic and superficial. Rigorous proofs and more precise details can be found in the books by Ferraty and Vieu [51], Horváth and Kokoszka [70], or Hsing and Eubank [71]. The notion of amplitude and phase variation is discussed in the books by Ramsay and Silverman [109, 110] that are of a more applied flavour. One can also consult the review by Wang et al. [127], where amplitude and phase variation are discussed in Sect. 5.2.

generally, for any finite set $x_1,\dots,x_n$, the $n\times n$ covariance matrix with entries $\kappa(x_i,x_j)=\mathrm{cov}[Y(x_i),Y(x_j)]$ encapsulates (up to second order) the stochastic deviations of the random vector $(Y(x_1),\dots,Y(x_n))$ from its mean, in analogy with the multivariate case. Heuristically, one then views amplitude variation as the collection $\kappa(x,y)$ for $x,y\in K$ in a sense we discuss next.

One typically views Y as a random element in the separable Hilbert space $L_2(K)$, assumed to have $\mathbb{E}\|Y\|^2<\infty$ and continuous sample paths, so that in particular $Y(x)$ is a random variable for all $x\in K$. Then the *mean function*

$$m(x)=\mathbb{E}Y(x),\quad x\in K$$

and the *covariance kernel*

$$\kappa(x,y)=\mathrm{cov}[Y(x),Y(y)],\quad x,y\in K$$

are well-defined and finite; we shall assume that they are continuous, which is equivalent to Y being *mean-square continuous*:

$$\mathbb{E}[Y(y)-Y(x)]^2\to 0,\qquad y\to x.$$

The covariance kernel κ gives rise to the *covariance operator* $\mathscr{R}:L_2(K)\to L_2(K)$, defined by

$$(\mathscr{R}f)(y)=\int_K\kappa(x,y)f(x)\,\mathrm{d}x,$$

a self-adjoint positive semidefinite Hilbert–Schmidt operator on $L_2(K)$. The justification to this terminology is the observation that when $m=0$, for all bounded $f,g\in L_2(K)$,

$$\mathbb{E}\left\langle Y,f\right\rangle\left\langle Y,g\right\rangle=\mathbb{E}\left[\int_{K^2}Y(x)f(x)Y(y)g(y)\,\mathrm{d}(x,y)\right]=\int_K g(y)(\mathscr{R}f)(y)\,\mathrm{d}y,$$

and so, without the restriction to $m=0$,

$$\mathrm{cov}\left[\left\langle Y,f\right\rangle,\left\langle Y,g\right\rangle\right]=\int_K g(y)(\mathscr{R}f)(y)\,\mathrm{d}y=\left\langle g,\mathscr{R}f\right\rangle.$$

The covariance operator admits an eigendecomposition $(r_k,\phi_k)_{k=1}^{\infty}$ such that $r_k\searrow 0$, $\mathscr{R}\phi_k=r_k\phi_k$ and (ϕ_k) is an orthonormal basis of $L_2(K)$. One then has the celebrated *Karhunen–Loève expansion*

$$Y(x)=m(x)+\sum_{k=1}^{\infty}\left\langle Y-m,\phi_k\right\rangle\phi_k(x)=m(x)+\sum_{k=1}^{\infty}\xi_k\phi_k(x).$$

A major feature in this expansion is the separation of the functional part from the stochastic part: the functions $\phi_k(x)$ are deterministic; the random variables ξ_k are scalars. This separation actually holds for any orthonormal basis; the role of choosing the eigenbasis of $\mathscr{R}$ is making ξ_k *uncorrelated*:

$$\mathrm{cov}(\xi_k, \xi_l) = \mathrm{cov}\left[\langle Y, \phi_k \rangle, \langle Y, \phi_l \rangle\right] = \langle \phi_l, \mathscr{R}\phi_k \rangle$$

vanishes when $k \neq l$ and equals r_k otherwise. For this reason, it is not surprising that using as ϕ_k the eigenfunctions yields the optimal representation of Y. Here, optimality is with respect to truncations: for any other basis (ψ_k) and any M,

$$\mathbb{E}\left\| Y - m - \sum_{k=1}^{M} \langle Y - m, \psi_k \rangle \psi_k \right\|^2 \geq \mathbb{E}\left\| Y - m - \sum_{k=1}^{M} \langle Y - m, \phi_k \rangle \phi_k \right\|^2$$

so that (ϕ_k) provides the best finite-dimensional approximation to Y. The approximation error on the right-hand side equals

$$\mathbb{E}\left\| \sum_{k=M+1}^{\infty} \xi_k \phi_k \right\|^2 = \sum_{k=M+1}^{\infty} r_k$$

and depends on how quickly the eigenvalues of $\mathscr{R}$ decay.

One carries out inference for m and κ on the basis of a sample $Y_1, \dots, Y_n$ by

$$\widehat{m}(x) = \frac{1}{n} \sum_{i=1}^{n} Y_i(x), \qquad x \in K$$

and

$$\widehat{\kappa}(x, y) = \frac{1}{n} \sum_{i=1}^{n} Y_i(x) Y_i(y) - \widehat{m}(x)\widehat{m}(y),$$

from which one proceeds to estimate $\mathscr{R}$ and its eigendecomposition.

We have seen that amplitude variation in the sense described above is linear and dealt with using linear operations. There is another, qualitatively different type of variation, *phase variation*, that is non-linear and does not have an obvious finite-dimensional analogue. It arises when in addition to the randomness in the values $Y(x)$ itself, an extra layer of stochasticity is present in its domain of definition. In mathematical terms, there is a random invertible *warp function* (sometimes called *deformation* or *warping*) $T : K \to K$ and instead of $Y(x)$, one observes realisations from

$$\tilde{Y}(x) = Y(T^{-1}(x)), \qquad x \in K.$$

For this reason, phase variation can be viewed as "variation in the x-axis". When $d = 1$, the set K is usually interpreted as a time interval, and then the model stipulates that each individual has its own time scale. Typically, the warp function is assumed to be a homeomorphism of K independent of Y and often some additional smoothness is imposed, say $T \in C^2$. One of the classical examples is growth curves of children, of which a dataset from the Berkeley growth study (Jones and Bayley [73]) is shown in Fig. 4.1. The curves are the derivatives of the height of a sample of ten girls as a function of time, from birth until age 18. One clearly notices the presence of the two types of variation in the figure. The initial velocity for all children is the highest immediately or shortly after birth, and in most cases decreases sharply during the first 2 years. Then follows a period of acceleration for another year or so, and so on. Despite presenting qualitatively similar behaviour, the curves differ substantially not only in the magnitude of the peaks but also in their location. For instance, one red curve has a local minimum at the age of three, while a green one has a local maximum at almost that same time point. It is apparent that if one tries to estimate the mean function by averaging the curves at each time x, the shape of the resulting estimate would look very different from each of the curves. Thus, this pointwise averaging (known as the *cross-sectional mean*) fails to represent the typical behaviour. This phenomenon is seen more explicitly in the next example. The terminology of amplitude and phase comes from trigonometric functions, from

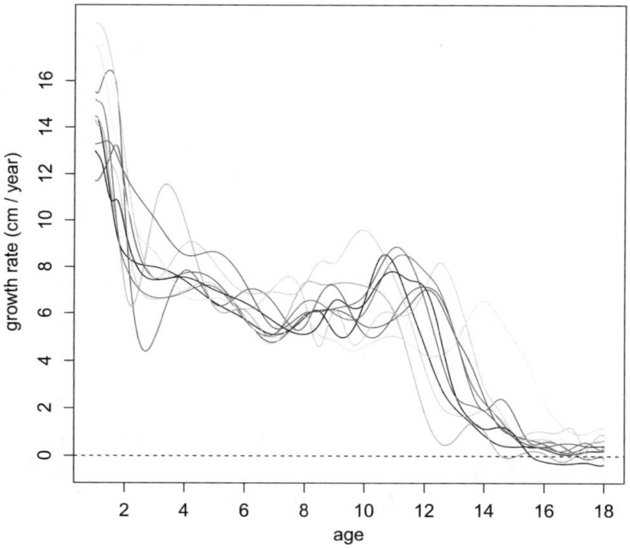

Fig. 4.1: Derivatives of growth curves of ten girls from the Berkeley dataset. The data and the code for the figure are from the R package fda (Ramsay et al. [111])

which we derive an artificial example that illustrates the difficulties of estimation in the presence of phase variation. Let A and B be symmetric random variables and consider the random function

$$\tilde{Y}(x) = A\sin[8\pi(x+B)]. \tag{4.1}$$

(Strictly speaking, $x \mapsto x+B$ is not from $[0,1]$ to itself; for illustration purposes, we assume in this example that $K = \mathbb{R}$.) The random variable A generates the amplitude variation, while B represents the phase variation. In Fig. 4.2, we plot four realisations and the resulting empirical means for the two extreme scenarios where $B = 0$ (no phase variation) or $A = 1$ (no amplitude variation). In the left panel of the figure, we see that the sample mean (in thick blue) lies between the observations and has a similar form, so can be viewed as the curve representing the typical realisation of the random curve. This is in contrast to the right panel, where the mean is qualitatively different from all curves in the sample: though periodicity is still present, the peaks and troughs have been flattened, and the sample mean is much more diffuse than any of the observations.

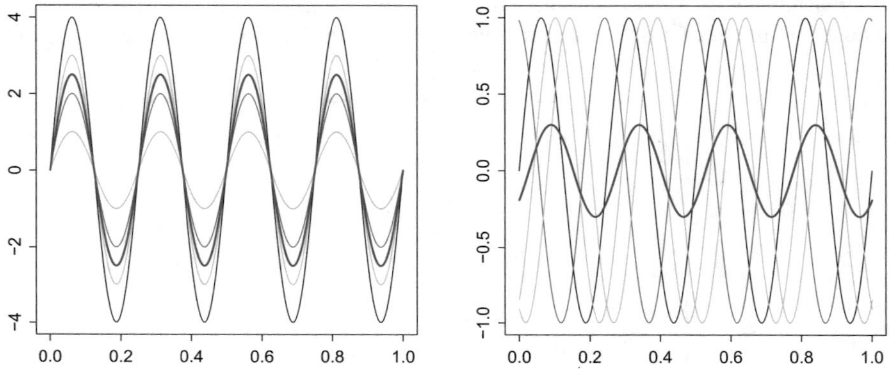

Fig. 4.2: Four realisations of (4.1) with means in thick blue. Left: amplitude variation ($B = 0$); right: phase variation ($A = 1$)

The phenomenon illustrated in Fig. 4.2 is hardly surprising, since as mentioned earlier amplitude variation is linear while phase variation is not, and taking sample means is a linear operation. Let us see in formulae how this phenomenon occurs. When $A = 1$ we have

$$\mathbb{E}\tilde{Y}(x) = \sin(8\pi x)\mathbb{E}[\cos(8\pi B)] + \cos(8\pi x)\mathbb{E}[\sin(8\pi B)].$$

Since B is symmetric the second term vanishes, and unless B is trivial the expectation of the cosine is smaller than one in absolute value. Consequently, the expectation of $\tilde{Y}(x)$ is the original function $\sin 8\pi x$ multiplied by a constant of magnitude strictly less than one, resulting in peaks of smaller magnitude.

In the general case, where $\tilde{Y}(x) = Y(T^{-1}(x))$ and Y and T are independent, we have

$$\mathbb{E}\tilde{Y}(x) = \mathbb{E}[m(T^{-1}(x))]$$

and

$$\mathrm{cov}[\tilde{Y}(x), \tilde{Y}(y)] = \mathbb{E}[\kappa(T^{-1}(x), T^{-1}(y))] + \mathrm{cov}[m(T^{-1}(x), m(T^{-1}(y))].$$

From this, several conclusions can be drawn. Let $\tilde{\mu} = \mu(T^{-1}(x))$ be the conditional mean function given T. Then the value of the mean function itself, $\mathbb{E}\tilde{\mu}$, at x_0 is determined not by a single point, say x, but rather by all the values of m at the possible outcomes of $T^{-1}(x)$. In particular, if x_0 was a local maximum for m, then $\mathbb{E}[\tilde{\mu}(x_0)]$ will typically be strictly smaller than $m(x_0)$; the phase variation results in smearing m.

At this point an important remark should be made. Whether or not phase variation is problematic depends on the specific application. If one is interested indeed in the mean and covariance functions of $\tilde{Y}$, then the standard empirical estimators will be consistent, since $\tilde{Y}$ itself is a random function. But if it is rather m, the mean of Y, that is of interest, then the confounding of the amplitude and phase variation will lead to inconsistency. This can also be seen from the formula

$$\tilde{Y}(x) = m(T^{-1}(x)) + \sum_{k=1}^{\infty} \xi_k \phi_k(T^{-1}(x)).$$

The above series is *not* the Karhunen–Loève expansion of $\tilde{Y}$; the simplest way to notice this is the observation that $\phi_k(T^{-1}(x))$ includes both the functional component ϕ_k and the random component $T^{-1}(x)$. The true Karhunen–Loève expansion of $\tilde{Y}$ will in general be qualitatively very different from that of Y, not only in terms of the mean function but also in terms of the covariance operator and, consequently, its eigenfunctions and eigenvalues. As illustrated in the trigonometric example, the typical situation is that the mean $\mathbb{E}\tilde{Y}$ is more diffuse than m, and the decay of the eigenvalues $\tilde{r}_k$ of the covariance operator is slower than that of r_k; as a result, one needs to truncate the sum at high threshold in order to capture a substantial enough part of the variability. In the toy example (4.1), the Karhunen–Loève expansion has a single term besides the mean if $B = 0$, while having two terms if $A = 1$.

When one is indeed interested in the mean m and the covariance κ, the random function T pertaining to the phase variation is a nuisance parameter. Given a sample $\tilde{Y}_i = Y_i \circ T_i^{-1}$, $i = 1, \dots, n$, there is no point in taking pointwise means of $\tilde{Y}_i$, because the curves are *misaligned*; $\tilde{Y}_1(x) = Y_1(T_1^{-1}(x))$ should not be compared with $\tilde{Y}_2(x)$, but rather with $Y_2(T_1^{-1}(x)) = \tilde{Y}_2(T_1^{-1}(T_2(x)))$. To overcome this difficulty, one seeks estimators $\widehat{T_i}$ such that

$$\widehat{Y_i}(x) = \tilde{Y}_i(\widehat{T_i}(x)) = Y_i(T_i^{-1}(\widehat{T_i}(x)))$$

is approximately $Y_i(x)$. In other words, one tries to align the curves in the sample to have a common time scale. Such a procedure is called *curve registration*. Once registration has been carried out, one proceeds the analysis on $\widehat{Y_i}(x)$ assuming only amplitude variation is now present: estimate the mean m by

$$\widehat{m}(x) = \frac{1}{n} \sum_{i=1}^{n} \widehat{Y_i}(x)$$

and the covariance κ by its analogous counterpart. Put differently, registering the curves amounts to *separating the two types of variation*. This step is crucial regardless of whether the warp functions are considered as nuisance or an analysis of the warp functions is of interest in the particular application.

There is an obvious identifiability problem in the model $\tilde{Y} = Y \circ T^{-1}$. If S is any (deterministic) invertible function, then the model with (Y,T) is statistically indistinguishable from the model with $(Y \circ S, T \circ S)$. It is therefore often assumed that $\mathbb{E}T = \mathbf{i}$ is the identity and in addition, in nearly all application, that T is monotonically increasing (if $d = 1$).

Discretely observed data. One cannot measure the height of person at every single instant of her life. In other words, it is rare in practice that one has access to the entire curve. A far more common situation is that one observes the curves *discretely*, i.e., at a finite number of points. The conceptually simplest setting is that one has access to a grid $x_1, \dots, x_J \in K$, and the data come in the form

$$\tilde{y}_{ij} = \tilde{Y}_i(t_j),$$

possibly with measurement error. The problem is to find, given $\tilde{y}_{ij}$, consistent estimators of T_i and of the original, aligned functions Y_i.

In the bibliographic notes, we review some methods for carrying out this separation of amplitude and phase variation. It is fair to say that no single registration method arises as the canonical solution to the functional registration problem. Indeed, most need to make additional structural and/or smoothness assumptions on the warp maps, further to the basic identifiability conditions requiring that T be increasing and that $\mathbb{E}T$ equal the identity. We will eventually see that, in contrast, the case of point processes (viewed as discretely observed random measures) admits a canonical framework, without needing additional assumptions.

4.1.2 The Point Process Case

A point process is the mathematical object that represents the intuitive notion of a random collection of points in a space $\mathscr{X}$. It is formally defined as a measurable map Π from a generic probability space into the space of (possibly infinite) Borel integer-valued measures of $\mathscr{X}$ in such a way that $\Pi(B)$ is a measurable real-valued random variable for all Borel subsets B of $\mathscr{X}$. The quantity $\Pi(B)$ represents the random number of points observed in the set B. Among the plethora of books on point processes, let us mention Daley and Vere-Jones [41] and Karr [79]. Kallenberg [75] treats more general objects, *random measures*, of which point processes are a peculiar special case. We will assume for convenience that Π is a measure on a compact subset $K \subset \mathbb{R}^d$.

Amplitude variation of Π can be understood in analogy with the functional case. One defines the mean measure

$$\lambda(A) = \mathbb{E}[\Pi(A)], \qquad A \subset K \text{ Borel}$$

and, provided that $\mathbb{E}[\Pi(K)]^2 < \infty$, the covariance measure

$$\kappa(A,B) = \mathrm{cov}[\Pi(A),\Pi(B)] = \mathbb{E}[\Pi(A)\Pi(B)] - \lambda(A)\lambda(B),$$

the latter being a finite signed Borel measure on K. Just like in the functional case, these two objects encapsulate the (second-order) amplitude variation[3] properties of the law of Π.

Given a sample $\Pi_1,\dots,\Pi_n$ of independent point processes distributed as Π, the natural estimators

$$\widehat{\lambda}(A) = \frac{1}{n}\sum_{i=1}^{n}\Pi_i(A); \qquad \widehat{\kappa}(A,B) = \frac{1}{n}\sum_{i=1}^{n}\Pi_i(A)\Pi_i(B) - \widehat{\lambda}(A)\widehat{\lambda}(B),$$

are consistent and the former asymptotically normal [79, Proposition 4.8].

Phase variation then pertains to a random warp function $T : K \to K$ (independent of Π) that deforms Π: if we denote the points of Π by $x_1,\dots,x_K$ (with K random), then instead of (x_i), one observes $T(x_1),\dots,T(x_K)$. In symbols, this means that the data arise as $\widetilde{\Pi} = T\#\Pi$. We refer to Π as the *original point processes*, and $\widetilde{\Pi}$ as the *warped point processes*. An example of 30 warped and unwarped point processes is shown in Fig. 4.3. The point patterns in both panels present a qualitatively similar structure: there are two peaks of high concentration of points, while few points appear between these peaks. The difference between the two panels is in the position and concentration of those peaks. In the left panel, only amplitude variation is present, and the location/concentration of the peaks is the same across all observations. In contrast, phase variation results in shifting the peaks to different places for each of the observations, while also smearing or sharpening them. Clearly, estimation of the mean measure of a subset A by averaging the number of observed points in A would not be satisfactory as an estimator of λ when carried out with the warped data. As in the functional case, it will only be consistent for the measure $\tilde{\lambda}$ defined by

$$\tilde{\lambda}(A) = \mathbb{E}[\lambda(T^{-1}(A))], \qquad A \subseteq \mathscr{X},$$

and $\tilde{\lambda} = \mathbb{E}[T\#\lambda]$ misses most (or at least a significant part) of the bimodal structure of λ and is far more diffuse.

[3] If the cumulative count process $\Gamma(t) = \Pi[0,t)$ is mean-square continuous, then the use of the term "amplitude variation" can be seen to remain natural, as $\Gamma(t)$ will admit a Karhunen–Loève expansion, with all stochasticity being attributable to the random amplitudes in the expansion.

Since Π and T are independent, the conditional expectation of $\widetilde{\Pi}$ given T is

$$\mathbb{E}[\widetilde{\Pi}(A)|T] = \mathbb{E}[\Pi(T^{-1}(A))|T] = \lambda(T^{-1}(A)) = [T\#\lambda](A).$$

Consequently, we refer to $\Lambda = T\#\lambda$ as the *conditional mean measure*. The problem of separation of amplitude and phase variation can now be stated as follows. On the basis of a sample $\widetilde{\Pi}_1, \dots, \widetilde{\Pi}_n$, find estimators of (T_i) and (Π_i). Registering the point processes amounts to constructing estimators, *registration maps* $\widehat{T_i^{-1}}$, such that the aligned points

$$\widehat{\Pi_i} = \widehat{T_i^{-1}}\#\widetilde{\Pi}_i = [\widehat{T_i^{-1}} \circ T_i]\#\Pi_i$$

are close to the original points Π_i.

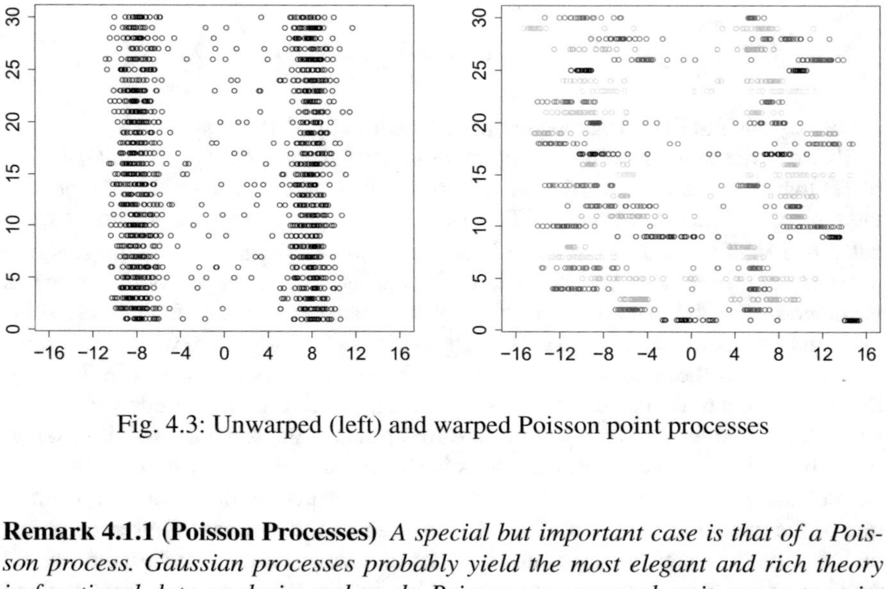

Fig. 4.3: Unwarped (left) and warped Poisson point processes

Remark 4.1.1 (Poisson Processes) *A special but important case is that of a Poisson process. Gaussian processes probably yield the most elegant and rich theory in functional data analysis, and so do Poisson processes when it comes to point processes. We say that Π is a* Poisson process *when the following two conditions hold. (1) For any disjoint collection $(A_1, \dots, A_n)$ of Borel sets, the random variables $\Pi(A_1), \dots, \Pi(A_n)$ are independent; and (2) for every Borel $A \subseteq K$, $\Pi(A)$ follows a Poisson distribution with mean $\lambda(A)$:*

$$\mathbb{P}(\Pi(A) = k) = e^{-\lambda(A)} \frac{[\lambda(A)]^k}{k!}.$$

Conditional upon T, the random variables $\widetilde{\Pi}(A_k) = \Pi(T^{-1}(A_k))$, $k = 1, \dots, n$ are independent as the sets $(T^{-1}(A_k))$ are disjoint, and $\widetilde{\Pi}(A)$ follows a Poisson distribution with mean $\lambda(T^{-1}(A)) = \Lambda(A)$. This is precisely the definition of a Cox process*: conditional upon the* driving measure Λ, *$\widetilde{\Pi}$ is a Poisson process with mean measure λ. For this reason, it is also called a* doubly stochastic process*; in our context, the* *phase variation is associated with the stochasticity of Λ while the amplitude one is associated with the Poisson variation conditional upon Λ.*

As in the functional case there are problems with identifiability: the model (Π, T) cannot be distinguished from the model $(S\#\Pi, T \circ S^{-1})$ for any invertible $S : K \to K$. It is thus natural to assume that $\mathbb{E}T$ is the identity map[4] (otherwise set $S = \mathbb{E}T$, i.e., replace Π by $[\mathbb{E}T]\#\Pi$ and T by $T \circ [\mathbb{E}T]^{-1}$).

Constraining T to have mean identity is nevertheless not sufficient for the model $\widetilde{\Pi} = T\#\Pi$ to be identifiable. The reason is that given the two point sets $\widetilde{\Pi}$ and Π, there are many functions that push forward the latter to the former. This ambiguity can be dealt with by assuming some sort of *regularity* or *parsimony* for T. For example, when $K = [a, b]$ is a subset of the real line, imposing T to be monotonically increasing guarantees its uniqueness. In multiple dimensions, there is no obvious analogue for increasing functions. One possible definition is the monotonicity described in Sect. 1.7.2:

$$\langle T(y) - T(x), y - x \rangle \geq 0, \qquad x, y \in K.$$

This property is rather weak in a sense we describe now. Let $K \subseteq \mathbb{R}^2$ and write $y \geq x$ if and only if $y_i \geq x_i$ for $i = 1, 2$. It is natural to expect the deformations to maintain the *lexicographic order* in $\mathbb{R}^2$:

$$y \geq x \quad \Longrightarrow \quad T(y) \geq T(x).$$

If we require in addition that the ordering must be preserved for all quadrants: for $z = T(x)$ and $w = T(y)$

$$\{y_1 \geq x_1, y_2 \leq x_2\} \qquad \Longrightarrow \{w_1 \geq z_1, w_2 \leq z_2\},$$

then monotonicity is automatically satisfied. In that sense, it is arguably not very restrictive.

Monotonicity is weaker than cyclical monotonicity (see (1.10) with $y_i = T(x_i)$), which is itself equivalent to the property of being the subgradient of a convex function. But if extra smoothness is present and T is a gradient of some function $\phi : K \to \mathbb{R}$, then ϕ must be convex and T is then cyclically monotone. Consequently, we will make the following assumptions:

- the expected value of T is the identity;
- T is a gradient of a convex function.

In the functional case, at least on the real line, these two conditions are imposed on the warp functions in virtually all applications, often accompanied with additional assumptions about smoothness of T, its structural properties, or its distance from the identity. In the next section, we show how these two conditions alone lead to the Wasserstein geometry and open the door to consistent, fully nonparametric separation of the amplitude and phase variation.

---

[4] This can be defined as Bochner integral in the space of measurable bounded $T : K \to K$.

4.2 Wasserstein Geometry and Phase Variation

4.2.1 Equivariance Properties of the Wasserstein Distance

A first hint to the relevance of Wasserstein metrics in $\mathcal{W}_p(\mathcal{X})$ for deformations of the space $\mathcal{X}$ is that for all $p \geq 1$ and all $x, y \in \mathcal{X}$,

$$W_p(\delta_x, \delta_y) = \|x - y\|,$$

where δ_x is as usual the Dirac measure at $x \in \mathcal{X}$. This is in contrast to metrics such as the bounded Lipschitz distance (that metrises weak convergence) or the total variation distance on $P(\mathcal{X})$. Recall that these are defined by

$$\|\mu - \nu\|_{\mathrm{BL}} = \sup_{\|\varphi\|_{\mathrm{BL}} \leq 1} \left| \int_{\mathcal{X}} \varphi \, \mathrm{d}\mu - \int_{\mathcal{X}} \varphi \, \mathrm{d}\nu \right|; \qquad \|\mu - \nu\|_{\mathrm{TV}} = \sup_A |\mu(A) - \nu(A)|,$$

so that

$$\|\delta_x - \delta_y\|_{\mathrm{BL}} = \min(1, \|x - y\|); \qquad \|\delta_x - \delta_y\|_{\mathrm{TV}} = \begin{cases} 1 & x \neq y \\ 0 & x = y. \end{cases}$$

In words, the total variation metric "does not see the geometry" of the space $\mathcal{X}$. This is less so for the bounded Lipschitz distance that does take small distances into account but not large ones.

Another property (shared by BL and TV) is equivariance with respect to translations. It is more convenient to state it using the probabilistic formalism of Sect. 1.2. Let $X \sim \mu$ and $Y \sim \nu$ be random elements in $\mathcal{X}$, a be a fixed point in $\mathcal{X}$, $X' = X + a$ and $Y' = Y + a$. Joint couplings $Z' = (X', Y')$ are precisely those that take the form $(a, a) + Z$ for a joint coupling $Z = (X, Y)$. Thus

$$W_p(\mu * \delta_a, \nu * \delta_a) = W_p(X' + a, Y' + a) = W_p(X, Y) = W_p(\mu, \nu),$$

where δ_a is a Dirac measure at a and $*$ denotes convolution.

This carries over to Fréchet means in an obvious way.

Lemma 4.2.1 (Fréchet Means and Translations) *Let Λ be a random measure in $\mathcal{W}_2(\mathcal{X})$ with finite Fréchet functional and $a \in \mathcal{X}$. Then γ is a Fréchet mean of Λ if and only if $\gamma * \delta_a$ is a Fréchet mean of $\Lambda * \delta_a$.*

The result holds for other values of p, in the formulation sketched in the bibliographic notes of Chap. 2. In the quadratic case, one has a simple extension to the case where only one measure is translated. Denote the first moment (mean) of $\mu \in \mathcal{W}_1(\mathcal{X})$ by

$$m : \mathcal{W}_1(\mathcal{X}) \to \mathcal{X} \qquad m(\mu) = \int_{\mathcal{X}} x \, \mathrm{d}\mu(x).$$

(When $\mathscr{X}$ is infinite-dimensional, this can be defined as the unique element $m \in \mathscr{X}$ satisfying

$$\langle m, y \rangle = \int_{\mathscr{X}} \langle x, y \rangle \, \mathrm{d}\mu(x), \qquad y \in \mathscr{X}.)$$

By an equivalence of couplings similar to above, we obtain

$$W_2^2(\mu * \delta_a, \nu) = W_2^2(\mu, \nu) + (a - [m(\mu) - m(\nu)])^2 - [m(\mu) - m(\nu)]^2,$$

which is minimised at $a = m(\mu) - m(\nu)$. This leads to the following conclusion:

Proposition 4.2.2 (First Moment of Fréchet Mean) *Let Λ be a random measure in $\mathscr{W}_2(\mathscr{X})$ with finite Fréchet functional and Fréchet mean γ. Then*

$$\int_{\mathscr{X}} x \, \mathrm{d}\gamma(x) = \mathbb{E}\left[\int_{\mathscr{X}} x \, \mathrm{d}\Lambda(x)\right].$$

4.2.2 Canonicity of Wasserstein Distance in Measuring Phase Variation

The purpose of this subsection is to show that the standard functional data analysis assumptions on the warp function T, having mean identity and being increasing, are equivalent to purely geometric conditions on T and the conditional mean measure $\Lambda = T\#\lambda$. Put differently, if one is willing to assume that $\mathbb{E}T = \mathbf{i}$ and that T is increasing, then one is led *unequivocally* to the problem of estimation of Fréchet means in the Wasserstein space $\mathscr{W}_2(\mathscr{X})$. When $\mathscr{X} \neq \mathbb{R}$, "increasing" is interpreted as being the gradient of a convex function, as explained at the end of Sect. 4.1.2.

The total mass $\lambda(\mathscr{X})$ is invariant under the push-forward operation, and when it is finite, we may assume without loss of generality that it is equal to one, because all the relevant quantities scale with the total mass. Indeed, if $\lambda = \tau\mu$ with μ probability measure and $\tau > 0$, then $T\#\lambda = \tau \times T\#\mu$, and the Wasserstein distance (defined as the infimum-over-coupling integrated cost) between $\tau\mu$ and $\tau\nu$ is $\tau W_p(\mu, \nu)$ for μ, ν probabilities.

We begin with the one-dimensional case, where the explicit formulae allow for a more transparent argument, and for simplicity we will assume some regularity.

Assumptions 2 *The domain $K \subset \mathbb{R}$ is a nonempty compact convex set (an interval), and the continuous and injective random map $T : K \to \mathbb{R}$ (a random element in $C_b(K)$) satisfies the following two conditions:*

(A1) Unbiasedness: $\mathbb{E}[T(x)] = x$ *for all* $x \in K$.
(A2) Regularity: T *is monotone increasing.*

The relevance of the Wasserstein geometry to phase variation becomes clear in the following proposition that shows that Assumptions 2 are equivalent to geometric assumptions on the Wasserstein space $\mathscr{W}_2(\mathbb{R})$.

Proposition 4.2.3 (Mean Identity Warp Functions and Fréchet Means in $\mathscr{W}_2(\mathbb{R})$) *Let $\phi \subset K \subset \mathbb{R}$ compact and convex and $T : K \to \mathbb{R}$ continuous. Then Assumptions 2 hold if and only if, for any $\lambda \in \mathscr{W}_2(K)$ supported on K such that $\mathbb{E}[W_2^2(T\#\lambda,\lambda)] < \infty$, the following two conditions are satisfied:*

(B1) Unbiasedness: *for any $\theta \in \mathscr{W}_2(\mathbb{R})$*

$$\mathbb{E}[W_2^2(T\#\lambda,\lambda)] \le \mathbb{E}[W_2^2(T\#\lambda,\theta)].$$

(B2) Regularity: *if $Q : K \to \mathbb{R}$ is such that $T\#\lambda = Q\#\lambda$, then with probability one*

$$\int_K \left|T(x)-x\right|^2 \mathrm{d}\lambda(x) \le \int_K \left|Q(x)-x\right|^2 \mathrm{d}\lambda(x), \qquad \textit{almost surely.}$$

These assumptions have a clear interpretation: (B1) stipulates that λ is a Fréchet mean of the random measure $\Lambda = T\#\lambda$, while (B2) states that T must be the optimal map from λ to Λ, that is, $T = \mathbf{t}_\lambda^\Lambda$.

Proof. If T satisfies (B2) then, as an optimal map, it must be nondecreasing λ-almost surely. Since λ is arbitrary, T must be nondecreasing on the entire domain K. Conversely, if T is nondecreasing, then it is optimal for any λ. Hence (A2) and (B2) are equivalent.

Assuming (A2), we now show that (A1) and (B1) are equivalent. Condition (B1) is equivalent to the assertion that for all $\theta \in \mathscr{W}_2(\mathbb{R})$,

$$\mathbb{E}\|F_{T\#\lambda}^{-1} - F_\lambda^{-1}\|_{L_2(0,1)}^2 = \mathbb{E}[W_2^2(T\#\lambda,\lambda)] \le \mathbb{E}[W_2^2(T\#\lambda,\theta)] = \mathbb{E}\|F_{T\#\lambda}^{-1} - F_\theta^{-1}\|_{L_2(0,1)}^2,$$

which is in turn equivalent to $\mathbb{E}[F_{T\#\lambda}]^{-1}] = \mathbb{E}[F_\Lambda^{-1}] = F_\lambda^{-1}$ (see Sect. 3.1.4). Condition (A2) and the assumptions on T imply that $F_\Lambda(x) = F_\lambda(T^{-1}(x))$. Suppose that F_λ is invertible (i.e., continuous and strictly increasing on K). Then $F_\Lambda^{-1}(u) = T(F_\lambda^{-1}(u))$. Thus (B1) is equivalent to $\mathbb{E}T(x) = x$ for all x in the range of F_λ^{-1}, which is K. The assertion that (A1) implies (B1), even if F_λ is not invertible, is proven in the next theorem (Theorem 4.2.4) in a more general context.

The situation in more than one dimension is similar but the proof is less transparent. To avoid compactness assumptions, we introduce the following power growth condition (taken from Agueh and Carlier [2]) of continuous functions that grow like $\|\cdot\|^q$ ($q \ge 0$):

$$G_q(\mathscr{X}) = (1+\|\cdot\|^q)C_b(\mathscr{X}) = \left\{ f : \mathscr{X} \to \mathbb{R} \text{ continuous} : \sup_{x\in\mathscr{X}} \frac{|f(x)|}{1+\|x\|^q} < \infty \right\}$$

with the norm $\|f\|_{G_q} = \sup|f(x)|/(1+\|x\|^q) = \|f/(1+\|\cdot\|^q)\|_\infty$. The space $G_q(\mathscr{X},\mathscr{X})$ is defined similarly, with f taking values in $\mathscr{X}$ instead of $\mathbb{R}$, and the norm will be denoted in the same way. These are nonseparable Banach spaces.

Theorem 4.2.4 (Mean Identity Warp Functions and Fréchet Means) *Fix $\lambda \in P(\mathscr{X})$ and let $\mathbf{t} \in G_1(\mathscr{X},\mathscr{X})$ be a (Bochner measurable) random optimal map* *with (Bochner) mean identity and such that* $\mathbb{E}\|\mathbf{t}\|_{G_1} < \infty$. *Then* $\Lambda = \mathbf{t}\#\lambda$ *has Fréchet mean* λ*:*

$$\mathbb{E}[W_2^2(\lambda,\Lambda)] \le \mathbb{E}[W_2^2(\theta,\Lambda)] \qquad \forall \theta \in \mathscr{W}_2(\mathscr{X}).$$

The generalisation with respect to the one-dimensional result is threefold. Firstly, since our main interest is the implication (A1–A2) $\Rightarrow$ (B1–B2), we need not assume T to be injective. Secondly, the support of λ is not required to be compact. Lastly, the result holds in arbitrary dimension, including infinite-dimensional separable Hilbert spaces $\mathscr{X}$. In particular, if $\mathbf{t}$ is a linear map, then $\|\mathbf{t}\|_{G_1}$ coincides with the operator norm of $\mathbf{t}$, so the assumption is that $\mathbf{t}$ be a bounded self-adjoint nonnegative operator with mean identity and finite expected operator norm.

Proof. Optimality of $\mathbf{t}$ ensures that it has a convex potential ϕ, and strong and weak duality give

$$W_2^2(\lambda,\Lambda) = \int_{\mathscr{X}} \left(\frac{1}{2}\|x\|^2 - \phi(x)\right) \mathrm{d}\lambda(x) + \int_{\mathscr{X}} \left(\frac{1}{2}\|y\|^2 - \phi^*(y)\right) \mathrm{d}\Lambda(y);$$

$$W_2^2(\theta,\Lambda) \ge \int_{\mathscr{X}} \left(\frac{1}{2}\|x\|^2 - \phi(x)\right) \mathrm{d}\theta(x) + \int_{\mathscr{X}} \left(\frac{1}{2}\|y\|^2 - \phi^*(y)\right) \mathrm{d}\Lambda(y).$$

Formally taking expectations, using Fubini's theorem and that $\mathbb{E}\phi = \|\cdot\|^2/2$ (since $\mathbb{E}\mathbf{t}$ is the identity) yields

$$\mathbb{E}[W_2^2(\theta,\Lambda)] \ge \int_{\mathscr{X}} \left(\frac{1}{2}\|x\|^2 - \mathbb{E}\phi(x)\right) \mathrm{d}\theta(x) + \mathbb{E}\left[\int_{\mathscr{X}} \left(\frac{1}{2}\|y\|^2 - \phi^*(y)\right) \mathrm{d}\Lambda(y)\right] = \mathbb{E}[W_2^2(\lambda,\Lambda)]$$

as required. The rigorous mathematical justification for this is given on page 88 in the supplement.

Remark 4.2.5 *The "natural" space for* $\mathbf{t}$ *would be* $\mathscr{L}_2(\lambda)$*, but without the continuity assumption, the result may fail (Álvarez-Esteban et al. [9, Example 3.1]). A simple argument shows that the growth condition imposed by the* G_1 *assumption is minimal; see page 89 in the supplement or Galasso et al. [58].*

Remark 4.2.6 *The same statement holds if* $\mathscr{X}$ *is replaced by a (Borel) convex subset* K *thereof. The integrals will then be taken on* K*, showing that* λ *minimises the Fréchet functional among measures supported on* K*, and, by continuity, on* $\overline{K}$*. By Proposition 3.2.4,* λ *is a Fréchet mean.*

4.3 Estimation of Fréchet Means

4.3.1 Oracle Case

In view of the canonicity of the Wasserstein geometry in Sect. 4.2.2, separation of amplitude and phase variation of the point processes $\widetilde{\Pi}_i$ essentially requires computing Fréchet means in the 2-Wasserstein space. It is both conceptually important and technically convenient to introduce the case where an oracle reveals the conditional mean measures $\Lambda = T\#\lambda$ entirely. Thus, assuming that $\lambda \in \mathscr{W}_2(\mathscr{X})$ is the unique Fréchet mean of a random measure Λ, the goal is to estimate the structural mean λ on the basis of independent and identically distributed realisations $\Lambda_1, \dots, \Lambda_n$ of λ.

Given that λ is defined as the minimiser of the Fréchet functional

$$F(\gamma) = \frac{1}{2}\mathbb{E}W_2^2(\Lambda, \gamma), \qquad \gamma \in \mathscr{W}_2(\mathscr{X}),$$

it is natural to estimate λ by a minimiser, say λ_n, of the empirical Fréchet functional

$$F_n(\gamma) = \frac{1}{2n}\sum_{i=1}^{n} W_2^2(\Lambda_i, \gamma), \qquad \gamma \in \mathscr{W}_2(\mathscr{X}).$$

A minimiser λ_n exists by Corollary 3.1.3. When $\mathscr{X} = \mathbb{R}$, λ_n can be seen to be an *unbiased* estimator of λ in a generalised sense of Lehmann [88] (see Sect. 4.3.5).

The warp maps (and their inverses) can then be estimated as the optimal maps from λ_n to each Λ_i (see Sect. 4.3.4).

4.3.2 Discretely Observed Measures

In practice, one does not have the fortune of fully observing the inherently infinite-dimensional objects $\Lambda_1, \dots, \Lambda_n$. A far more realistic scenario is that one only has access to a discrete version of Λ_i, say $\widetilde{\Lambda}_i$. The simplest situation is when $\widetilde{\Lambda}_i$ arises as an empirical measure of the form $\tau^{-1}\sum_{i=1}^{\tau}\delta\{Y_j\}$, where Y_j are independent with distribution Λ_i. More generally, $\widetilde{\Lambda}_i$ can be a normalised point process $\widetilde{\Pi}_i$ with mean measure $\tau\Lambda_i$, i.e.

$$\widetilde{\Lambda}_i = \frac{1}{\widetilde{\Pi}_i(\mathscr{X})}\widetilde{\Pi}_i \quad \text{with} \quad \mathbb{E}[\widetilde{\Pi}_i(A)|\Lambda_i] = \tau\Lambda_i(A), \qquad A \subseteq \mathscr{X} \text{ Borel}.$$

This encapsulates the case of empirical measure when τ is an integer and $\widetilde{\Pi}_i$ is a *binomial point process*. The parameter τ is the expected number of observed points over the entire space $\mathscr{X}$; the larger τ is, the more information $\widetilde{\Pi}_i$ gives on Λ_i.

Except if $\widetilde{\Lambda}_i$ is an empirical measure, there is one difficulty in the above setting that needs to be addressed. Unless $\widetilde{\Pi}_i$ is binomial, there is a positive probability that $\widetilde{\Pi}_i(\mathscr{X}) = 0$ and no points pertaining to Λ_i are observed. In the asymptotic setup below, conditions will be imposed to ensure that this probability becomes negligible as $n \to \infty$. For concreteness we define $\widetilde{\Lambda}_i = \lambda^{(0)}$ for some fixed measure $\lambda^{(0)}$ that will be of minor importance. This can be a Dirac measure at 0, a certain fixed Gaussian measure, or (normalised) Lebesgue measure on some bounded set in case $\mathscr{X} = \mathbb{R}^d$. We can now replace the estimator λ_n by $\widetilde{\lambda}_n$, defined as any minimiser of

$$\widetilde{F}_n(\gamma) = \frac{1}{2n} \sum_{i=1}^{n} W_2^2(\widetilde{\Lambda}_i, \gamma), \qquad \gamma \in \mathscr{W}_2(\mathscr{X}),$$

which exists by Corollary 3.1.3.

As a generalisation of the discrete case discussed in Sect. 1.3, the Fréchet mean of discrete measures can be computed exactly. Suppose that $N_i = \widetilde{\Pi}_i(\mathscr{X})$ is nonzero for all i. Then each $\widetilde{\Lambda}_i$ is a discrete measure supported on N_i points. One can then recast the multimarginal formulation (see Sect. 3.1.2) as a finite linear program, solve it, and "average" the solution as in Proposition 3.1.2 in order to obtain $\widetilde{\lambda}_n$ (an alternative linear programming formulation for finding a Fréchet mean is given by Anderes et al. [14]). Thus, $\widetilde{\lambda}_n$ can be computed in finite time, even when $\mathscr{X}$ is infinite-dimensional.

Finally, a remark about measurability is in order. Point processes can be viewed as random elements in $M_+(\mathscr{X})$ endowed with the *vague topology* induced from convergence of integrals of continuous functions with compact support. If μ_n converge to μ vaguely, and a_n are numbers that converge to a, then $a_n\mu_n \to a\mu$ vaguely. Thus, $\tilde{\Lambda}_i$ is a continuous function of the pair $(\widetilde{\Pi}_i, \widetilde{\Pi}_i(\mathscr{X}))$ and can be viewed as a random measure with respect to the vague topology. The restriction of the vague topology to probability measures is equivalent to the weak topology,[5] and therefore vague, weak, and Wasserstein measurability are all equivalent.

4.3.3 Smoothing

Even when the computational complexity involved in calculating $\widetilde{\lambda}_n$ is tractable, there is another reason not to use it as an estimator for λ. If one has a-priori knowledge that λ is smooth, it is often desirable to estimate it by a smooth measure. One way to achieve this would be to apply some smoothing technique to $\widetilde{\lambda}_n$ using, e.g., kernel density estimation. However, unless the number of observed points from each measure is the same $N_1 = \cdots = N_n = N$, $\widetilde{\lambda}_n$ will usually be concentrated on many points, essentially $N_1 + \cdots + N_n$ of them. In other words, the Fréchet mean is concentrated on many more points than each of the measures $\widetilde{\Lambda}_i$, thus potentially hindering its usefulness as a mean because it will not be a representative of the sample.

This is most easily seen when $\mathscr{X} = \mathbb{R}$, in which case each $\widetilde{\Lambda}_i$ is a discrete uniform measure on points $x_1^i < x_2^i < \cdots < x_{N_i}^i$, where we assume for simplicity that the points are not repeated (this will happen with probability one if Λ_i is diffuse). If we now set G_i to be the distribution function of $\widetilde{\Lambda}_i$, then the quantile function G_i^{-1} is piecewise constant on each interval $(k, k+1]/N_i$ with jumps at

$$G_i^{-1}(k/N_i) = x_k^i, \qquad k = 1, 2, \ldots, N_i.$$

[5] In finite dimensional (or more generally, locally compact metric) spaces. If $\mathscr{X}$ is an infinite-dimensional Hilbert space, the vague topology is trivial. This is stated and proved as Lemma 5 on page 27 in the supplement.

The Fréchet mean has quantile function $G^{-1}(u) = n^{-1}\sum G_i^{-1}(u)$ and will have jumps at every point of the form k/N_i for $k \leq N_i$ and $i = 1,\ldots,n$. In the worst-case scenario, when no pair from N_i has a common divisor, there will be

$$\left(\sum_{i=1}^{n} N_i - 1\right) + 1 = \left(\sum_{i=1}^{n} N_i\right) - n + 1$$

jumps for G^{-1}, which is the number of points on which the Fréchet mean will be supported. (All the G_i^{-1}'s have a jump at one which thus needs to be counted once rather than n times.)

By counting the number of redundancies in the constraints matrix of the linear program, one can show that this is in general an upper bound on the number of support points of the Fréchet mean.

An alternative approach is to first smooth each observation $\widetilde{\lambda}_n$ and then calculate the Fréchet mean. Since it is easy to bound the Wasserstein distances when dealing with convolutions, we will employ kernel density estimation, although other smoothing approaches could be used as well.

To simplify the exposition, we provide the technical details only when $\mathscr{X} = \mathbb{R}^d$, but a similar construction will work when the dimension of $\mathscr{X}$ is infinite. Let $\psi : \mathbb{R}^d \to (0,\infty)$ be a continuous, bounded, strictly positive isotropic density function with unit variance: $\psi(x) = \psi_1(\|x\|)$ with ψ_1 nonincreasing and

$$\int_{\mathbb{R}^d} \|x\|^2 \psi(x)\,\mathrm{d}x = 1 = \int_{\mathbb{R}^d} \psi(x)\,\mathrm{d}x.$$

(Besides the boundedness all these properties can be relaxed, and if $\mathscr{X} = \mathbb{R}$ even boundedness is not necessary.) A classical example for ψ is the standard Gaussian density in $\mathbb{R}^d$. Define the rescaled version $\psi_\sigma(x) = \sigma^{-d}\psi(x/\sigma)$ for all $\sigma > 0$. We can then replace $\widetilde{\Lambda}_i$ by a smooth proxy $\widetilde{\Lambda}_i * \psi_\sigma$. If $\widetilde{\Lambda}_i$ is a sum of Dirac masses at $x_1,\ldots,x_{N_i}$, then

$$\widetilde{\Lambda}_i * \psi_\sigma \quad \text{has density} \quad g(x) = \frac{1}{N_i}\sum_{j=1}^{N_i} \psi_\sigma(x - x_i).$$

If $N_i = 0$, one can either use $\lambda^{(0)}$ or $\lambda^{(0)} * \psi_\sigma$; this event will have negligible probability anyway.

For the purpose of approximating $\widetilde{\Lambda}_i$, this convolution is an acceptable estimator, because as was seen in the proof of Theorem 2.2.7,

$$W_2^2(\widetilde{\Lambda}_i, \widetilde{\Lambda}_i * \psi_\sigma) \leq \sigma^2.$$

But the measure $\widetilde{\Lambda}_i$ has a strictly positive density throughout $\mathbb{R}^d$. If we know that Λ is supported on a convex compact $K \subset \mathbb{R}^d$, it is desirable to construct an estimator that has the same support K. The first idea that comes to mind is to project $\widetilde{\Lambda}_i * \psi_\sigma$ to K (see Proposition 3.2.4), as this will further decrease the Wasserstein distance, but the resulting measure will then have positive mass on the boundary of K, and will not be absolutely continuous. We will therefore use a different strategy: eliminate all the mass outside K and redistribute it on K. The simplest way to do this is to restrict $\widetilde{\Lambda}_i * \psi_\sigma$ to K and renormalise the restriction to be a probability measure. For technical reasons, it will be more convenient to bound the Wasserstein distance when the restriction and renormalisation is done separately on each point of $\widetilde{\Lambda}_i$. This yields the measure

$$\widehat{\Lambda}_i = \frac{1}{N_i} \sum_{j=1}^{N_i} \left. \frac{\delta\{x_j\} * \psi_\sigma}{[\delta\{x_j\} * \psi_\sigma](K)} \right|_K, \tag{4.2}$$

Lemma 4.4.2 below shows that $W_2^2(\widetilde{\Lambda}_i, \widehat{\Lambda}_i) \leq C\sigma^2$ for some finite constant C. It is apparent that $\widehat{\Lambda}_i$ is a continuous function of $\widetilde{\Lambda}_i$ and σ, so $\widehat{\Lambda}_i$ is measurable; in any case this is not particularly important because σ will vanish, so $\widehat{\Lambda}_i = \widetilde{\Lambda}_i$ asymptotically and the latter is measurable.

Our final estimator $\widehat{\lambda}_n$ for λ is defined as the minimiser of

$$\widehat{F}_n(\gamma) = \frac{1}{2n} \sum_{i=1}^{n} W_2^2(\widehat{\Lambda}_i, \gamma), \qquad \gamma \in \mathscr{W}_2(\mathscr{X}).$$

Since the measures $\widehat{\Lambda}_i$ are absolutely continuous, $\widehat{\lambda}_n$ is unique. We refer to $\widehat{\lambda}_n$ as the *regularised Fréchet–Wasserstein estimator*, where the regularisation comes from the smoothing and the possible restriction to K.

In the case $\mathscr{X} = \mathbb{R}$, $\widehat{\lambda}_n$ can be constructed via averaging of quantile functions. Let $\widehat{G}_i$ be the distribution function of $\widehat{\Lambda}_i$. Then $\widehat{\lambda}_n$ is the measure with quantile function

$$F_{\widehat{\lambda}_n}^{-1}(u) = \frac{1}{n} \sum_{i=1}^{n} \widehat{G}_i^{-1}(u), \qquad u \in (0,1),$$

and distribution function

$$F_{\widehat{\lambda}_n}(x) = [F_{\widehat{\lambda}_n}^{-1}]^{-1}(x).$$

By construction, the $\widehat{G}_i$ are continuous and strictly increasing, so the inverses are proper inverses and one does not to use the right-continuous inverse as in Sect. 3.1.4.

If $\mathscr{X} = \mathbb{R}^d$ and $d \geq 2$, then there is no explicit expression for $\widehat{\lambda}_n$, although it exists and is unique. In the next chapter, we present a steepest descent algorithm that approximately constructs $\widehat{\lambda}_n$ by taking advantage of the differentiability properties of the Fréchet functional $\widehat{F}_n$ in Sect. 3.1.6.

4.3.4 Estimation of Warpings and Registration Maps

Once estimators $\widehat{\Lambda}_i$, $i = 1, \ldots, n$ and $\widehat{\lambda}_n$ are constructed, it is natural to estimate the map $T_i = \mathbf{t}_\lambda^{\Lambda_i}$ and its inverse $T_i^{-1} = \mathbf{t}_{\Lambda_i}^{\lambda}$ (when Λ_i are absolutely continuous; see the discussion after Assumptions 3 below) by the plug-in estimators

$$\widehat{T}_i = \mathbf{t}_{\widehat{\lambda}_n}^{\widehat{\Lambda}_i}, \qquad \widehat{T_i^{-1}} = (\widehat{T}_i)^{-1} = \mathbf{t}_{\widehat{\Lambda}_i}^{\widehat{\lambda}_n}.$$

The latter, the registration maps, can then be used in order to register the points Π_i via

$$\widehat{\Pi_i^{(n)}} = \widehat{T_i^{-1}} \# \widetilde{\Pi}_i^{(n)} = \left[\widehat{T_i^{-1}} \circ T_i\right] \# \Pi_i^{(n)}.$$

It is thus reasonable to expect that if $\widehat{T_i^{-1}}$ is a good estimator, then its composition with T_i should be close to the identity and $\widehat{\Pi}_i$ should be close to Π_i.

4.3.5 Unbiased Estimation When $\mathscr{X} = \mathbb{R}$

In the same way, Fréchet means extend the notion of mean to non-Hilbertian spaces, they also extend the definition of unbiased estimators. Let H be a separable Hilbert space (or a convex subset thereof) and suppose that $\widehat{\theta}$ is a random element in H whose distribution μ_θ depends on a parameter $\theta \in H$. Then $\widehat{\theta}$ is *unbiased* for θ if for all $\theta \in H$

$$\mathbb{E}_\theta \widehat{\theta} = \int_H x \, \mathrm{d}\mu_\theta(x) = \theta.$$

(We use the standard notation $\mathbb{E}_\theta g(\widehat{\theta}) = \int g(x) \, \mathrm{d}\mu_\theta(x)$ in the sequel.) This is equivalent to

$$\mathbb{E}_\theta \|\theta - \widehat{\theta}\|^2 \le \mathbb{E}_\theta \|\gamma - \widehat{\theta}\|^2, \qquad \forall \theta, \gamma \in H.$$

In view of that, one can define unbiased estimators of $\lambda \in \mathscr{W}_2$ as measurable functions $\delta = \delta(\Lambda_1, \dots, \Lambda_n)$ for which

$$\mathbb{E}_\lambda W_2^2(\lambda, \delta) \le \mathbb{E}_\lambda W_2^2(\gamma, \delta), \qquad \forall \gamma, \theta \in \mathscr{W}_2.$$

This definition was introduced by Lehmann [88].

Unbiased estimators allow us to avoid the problem of over-registering (the so-called pinching effect; Kneip and Ramsay [82, Section 2.4]; Marron et al. [90, p. 476]). An extreme example of over-registration is if one “aligns” all the observed patterns into a single fixed point x_0. The registration will then seem “successful” in the sense of having no residual phase variation, but the estimation is clearly biased because the points are not registered to the correct reference measure. Thus, requiring the estimator to be unbiased is an alternative to penalising the registration maps.

Due to the Hilbert space embedding of $\mathscr{W}_2(\mathbb{R})$, it is possible to characterise unbiased estimators in terms of a simple condition on their quantile functions. As a corollary, λ_n, the Fréchet mean of $\{\Lambda_1, \dots, \Lambda_n\}$, is unbiased. Our regularised Fréchet–Wasserstein estimator $\widehat{\lambda}_n$ can then be interpreted as *approximately unbiased*, since it approximates the unobservable λ_n.

Proposition 4.3.1 (Unbiased Estimators in $\mathscr{W}_2(\mathbb{R})$) *Let Λ be a random measure in $\mathscr{W}_2(\mathbb{R})$ with finite Fréchet functional and let λ be the unique Fréchet mean of Λ (Theorem 3.2.11). An estimator δ constructed as a function of a sample $(\Lambda_1,\dots,\Lambda_n)$ is unbiased for λ if and only if the left-continuous representatives (in $L_2(0,1)$) satisfy $\mathbb{E}[F_\delta^{-1}(x)] = F_\lambda^{-1}(x)$ for all $x \in (0,1)$.*

Proof. The proof is straightforward from the definition: δ is unbiased if and only if for all λ and all γ,

$$\mathbb{E}_\lambda \|F_\lambda^{-1} - F_\delta^{-1}\|_{L_2}^2 \le \mathbb{E}_\lambda \|F_\gamma^{-1} - F_\delta^{-1}\|_{L_2}^2,$$

which is equivalent to $\mathbb{E}_\lambda[F_\delta^{-1}] = F_\lambda^{-1}$. In other words, these two functions must equal almost everywhere on $(0,1)$, and their left-continuous representatives must equal everywhere (the fact that $\mathbb{E}_\lambda[F_\delta^{-1}]$ has such a representative was established in Sect. 3.1.4).

To show that $\delta = \lambda_n$ is unbiased, we simply invoke Theorem 3.2.11 twice to see that

$$\mathbb{E}[F_\delta^{-1}] = \mathbb{E}\left[\frac{1}{n}\sum_{i=1}^n F_{\Lambda_i}^{-1}\right] = \mathbb{E}[F_\Lambda^{-1}] = F_\lambda^{-1},$$

which proves unbiasedness of δ.

4.4 Consistency

In functional data analysis, one often assumes that the number of curves n and the number of observed points per curve m both diverge to infinity. An analogous framework for point processes would similarly require the number of point processes n as well as the expected number of points τ per processes to diverge. A technical complication arises, however, because the mean measures do not suffice to characterise the distribution of the processes. Indeed, if one is given a point processes Π with mean measure λ (not necessarily a probability measure), and τ is an integer, there is no unique way to define a process $\Pi^{(\tau)}$ with mean measure $\tau\lambda$. One can define $\Pi^{(\tau)} = \tau\Pi$, so that every point in Π will be counted τ times. Such a construction, however, can never yield a consistent estimator of λ, even when $\tau \to \infty$.

Another way to generate a point process with mean measure $\tau\lambda$ is to take a superposition of τ independent copies of Π. In symbols, this means

$$\Pi^{(\tau)} = \Pi_1 + \cdots + \Pi_\tau,$$

with (Π_i) independent, each having the same distribution as Π. This superposition scheme gives the possibility to use the law of large numbers. If τ is not an integer, then this construction is not well-defined but can be made so by assuming that the distribution of Π is *infinitely divisible*. The reader willing to assume that τ is always an integer can safely skip to Sect. 4.4.1; all the main ideas are developed first for integer values of τ and then extended to the general case.

A point process Π is infinitely divisible if for every integer m there exists a collection of m independent and identically distributed $\Pi_i^{(1/m)}$ such that

$$\Pi = \Pi_1^{(1/m)} + \cdots + \Pi_m^{(1/m)} \qquad \text{in distribution.}$$

If Π is infinitely divisible and $\tau = k/m$ is rational, then can define $\pi^{(\tau)}$ using km independent copies of $\Pi^{(1/m)}$:

$$\Pi^{(\tau)} = \sum_{i=1}^{km} \Pi_i^{(1/m)}.$$

One then deals with irrational τ via duality and continuity arguments, as follows. Define the *Laplace functional* of Π by

$$L_\Pi(f) = \mathbb{E}\left[e^{-\Pi f}\right] = \mathbb{E}\left[\exp\left(-\int_{\mathscr{X}} f \, \mathrm{d}\Pi\right)\right] \in [0,1], \qquad f : \mathscr{X} \to \mathbb{R}_+ \quad \text{Borel.}$$

The Laplace functional characterises the distribution of the point process, generalising the notion of Laplace transform of a random variable or vector (Karr [79, Theorem 1.12]). By definition, it translates convolutions into products. When $\Pi = \Pi^{(1)}$ is infinitely divisible, the Laplace functional L_1 of Π takes the form (Kallenberg [75, Chapter 6]; Karr [79, Theorem 1.43])

$$L_1(f) = \mathbb{E}\left[e^{-\Pi^{(1)} f}\right] = \exp\left[-\int_{M_+(\mathscr{X})} (1 - e^{-\mu f}) \, \mathrm{d}\rho(\mu)\right] \quad \text{for some } \rho \in M_+(M_+(\mathscr{X})).$$

The Laplace functional of $\Pi^{(\tau)}$ is $L_\tau(f) = [L_1(f)]^\tau$ for any rational τ, which simply amounts to multiplying the measure ρ by the scalar τ. One can then do the same for an irrational τ, and the resulting Laplace functional determines the distribution of $\Pi^{(\tau)}$ for all $\tau > 0$.

4.4.1 Consistent Estimation of Fréchet Means

We are now ready to define our asymptotic setup. The following assumptions will be made. Notice that the Wasserstein geometry does not appear explicitly in these assumptions, but is rather *derived* from them in view of Theorem 4.2.4. The compactness requirement can be relaxed under further moment conditions on λ and the point process Π; we focus on the compact case for the simplicity and because in practice the point patterns will be observed on a bounded observation window.

Assumptions 3 *Let $K \subset \mathbb{R}^d$ be a compact convex nonempty set, λ an absolutely continuous probability measure on K, and τ_n a sequence of positive numbers. Let Π be a point processes on K with mean measure λ. Finally, define $U = \mathrm{int}K$.*

- *For every n, let $\{\Pi_1^{(n)}, \dots, \Pi_n^{(n)}\}$ be independent point processes, each having the same distribution as a superposition of τ_n copies of Π.*
- *Let T be a random injective function on K (viewed as a random element in $C_b(K,K)$ endowed with the supremum norm) such that $T(x) \in U$ for $x \in U$ (that is, $T \in C_b(U,U)$) with nonsingular derivative $\nabla T(x) \in \mathbb{R}^{d\times d}$ for almost all $x \in U$, that is a gradient of a convex function. Let $\{T_1, \dots, T_n\}$ be independent and identically distributed as T.*
- *For every $x \in U$, assume that $\mathbb{E}[T(x)] = x$.*
- *Assume that the collections $\{T_n\}_{n=1}^{\infty}$ and $\{\Pi_i^{(n)}\}_{i\le n, n=1,2,\dots}$ are independent.*
- *Let $\widetilde{\Pi}_i^{(n)} = T_i\#\Pi_i^{(n)}$ be the warped point processes, having* conditional mean measures $\Lambda_i = T_i\#\lambda = \tau_n^{-1}\mathbb{E}\left\{\widetilde{\Pi}_i^{(n)} \middle| T_i\right\}$.
- *Define $\widehat{\Lambda}_i$ by the smoothing procedure* (4.2)*, using bandwidth $\sigma_i^{(n)} \in [0,1]$ (possibly random).*

The dependence of the estimators on n will sometimes be tacit. But Λ_i does not depend on n.

By virtue of Theorem 4.2.4, λ is a Fréchet mean of the random measure $\Lambda = T\#\lambda$. Uniqueness of this Fréchet mean will follow from Proposition 3.2.7 if we show that Λ is absolutely continuous with positive probability. This is indeed the case, since T is injective and has a nonsingular Jacobian matrix; see Ambrosio et al. [12, Lemma 5.5.3]. The Jacobian assumption can be removed when $\mathscr{X} = \mathbb{R}$, because Fréchet means are always unique by Theorem 3.2.11.

Notice that there is no assumption about the dependence between rows. Assumptions 3 thus cover, in particular, two different scenarios:

- *Full independence*: here the point processes are independent across rows, that is, $\Pi_i^{(n)}$ and $\Pi_i^{(n+1)}$ are also independent.
- *Nested observations*: here $\Pi_i^{(n+1)}$ includes the same points as $\Pi_i^{(n)}$ and additional points, that is, $\Pi_i^{(n+1)}$ is a superposition of $\Pi_i^{(n)}$ and another point process distributed as $(\tau_{n+1} - \tau_n)\Pi$.

Needless to say, Assumptions 3 also encompass binomial processes when τ_n are integers, as well as Poisson processes or, more generally, Poisson cluster processes.

We now state and prove the consistency result for the estimators of the conditional mean measures Λ_i and the structural mean measure λ.

Theorem 4.4.1 (Consistency) *If Assumptions 3 hold, $\sigma_n = n^{-1}\sum_{i=1}^n \sigma_i^{(n)} \to 0$ almost surely and $\tau_n \to \infty$ as $n \to \infty$, then:*

1. *The estimators $\widehat{\Lambda}_i$ defined by* (4.2)*, constructed with bandwidth $\sigma = \sigma_i^{(n)}$, are Wasserstein-consistent for the conditional mean measures: for all i such that $\sigma_i^{(n)} \xrightarrow{p} 0$*

$$W_2\left(\widehat{\Lambda}_i, \Lambda_i\right) \xrightarrow{p} 0, \qquad as\ n \to \infty;$$

2. *The regularised Fréchet–Wasserstein estimator of the structural mean measure (as described in Sect. 4.3) is strongly Wasserstein-consistent,*

$$W_2(\widehat{\lambda}_n, \lambda) \xrightarrow{a.s.} 0, \qquad as\ n \to \infty.$$

Convergence in 1. holds almost surely under the additional conditions that $\sum_{n=1}^{\infty} \tau_n^{-2} < \infty$ *and* $\mathbb{E}\left[\Pi(\mathbb{R}^d)\right]^4 < \infty$. *If* $\sigma_n \to 0$ *only in probability, then convergence in 2. still holds in probability.*

Theorem 4.4.1 still holds without smoothing ($\sigma_n = 0$). In that case, $\widehat{\lambda}_n = \tilde{\lambda}_n$ is possibly not unique, and the theorem should be interpreted in a set-valued sense (as in Proposition 1.7.8): almost surely, *any* choice of minimisers $\tilde{\lambda}_n$ converges to λ as $n \to \infty$.

The preceding paragraph notwithstanding, we will usually assume that some smoothing *is* present, in which case $\widehat{\lambda}_n$ is unique and absolutely continuous by Proposition 3.1.8. The uniform Lipschitz bounds for the objective function show that if we restrict the relevant measures to be absolutely continuous, then $\widehat{\lambda}_n$ is a continuous function of $(\widehat{\Lambda_1}, \dots, \widehat{\Lambda_n})$ and hence $\widehat{\lambda}_n : (\Omega, \mathscr{F}, \mathbb{P}) \to \mathscr{W}_2(K)$ is measurable; this is again a minor issue because many arguments in the proof hold for each $\omega \in \Omega$ separately. Thus, even if $\widehat{\lambda}_n$ is not measurable, the proof shows that the convergence holds outer almost surely or in outer probability.

The first step in proving consistency is to show that the Wasserstein distance between the unsmoothed and the smoothed estimators of Λ_i vanishes with the smoothing parameter. The exact rate of decay will be important to later establish the rate of convergence of $\widehat{\lambda}_n$ to λ, and is determined next.

Lemma 4.4.2 (Smoothing Error) *There exists a finite constant* $C_{\psi,K}$, *depending only on* ψ *and on* K, *such that*

$$W_2^2\left(\widehat{\Lambda}_i, \widetilde{\Lambda}_i\right) \le C_{\psi,K}\sigma^2 \quad if\ \sigma \le 1. \tag{4.3}$$

Since the smoothing parameter will anyway vanish, this restriction to small values of σ is not binding. The constant $C_{\psi,K}$ is explicit. When $\mathscr{X} = \mathbb{R}$, a more refined construction allows to improve this constant in some situations, see Panaretos and Zemel [100, Lemma 1].

Proof. The idea is that (4.2) is a sum of measures with mass $1/N_i$ that can be all sent to the relevant point x_j, and we refer to page 98 in the supplement for the precise details.

Proof (Proof of Theorem 4.4.1). The proof, detailed on page 97 of the supplement, follows the following steps: firstly, one shows the convergence in probability of $\widehat{\Lambda}_i$ to Λ_i. This is basically a corollary of Karr [79, Proposition 4.8] and the smoothing bound from Lemma 4.4.2.

To prove claim (2) one considers the functionals, defined on $\mathscr{W}_2(K)$:

$$F(\gamma) = \frac{1}{2}\mathbb{E}W_2^2(\Lambda,\gamma);$$

$$F_n(\gamma) = \frac{1}{2n}\sum_{i=1}^n W_2^2(\Lambda_i,\gamma);$$

$$\tilde{F}_n(\gamma) = \frac{1}{2n}\sum_{i=1}^n W_2^2(\widetilde{\Lambda}_i,\gamma), \qquad \widetilde{\Lambda}_i = \frac{\widetilde{\Pi}_i^{(n)}}{N_i^{(n)}} \quad \text{or } \lambda^{(0)} \text{ if } N_i^{(n)} = 0;$$

$$\widehat{F}_n(\gamma) = \frac{1}{2n}\sum_{i=1}^n W_2^2(\widehat{\Lambda}_i,\gamma), \qquad \widehat{\Lambda}_i = \lambda^{(0)} \text{ if } N_i^{(n)} = 0.$$

Since K is compact, they are all locally Lipschitz, so their differences can be controlled by the distances between Λ_i, $\widetilde{\Lambda}_i$, and $\widehat{\Lambda}_i$. The first distance vanishes since the intensity $\tau \to \infty$, and the second by the smoothing bound. Another compactness argument yields that $\widehat{F}_n \to F$ uniformly on $\mathscr{W}_2(K)$, and so the minimisers converge.

The almost sure convergence in (1) is proven as follows. Under the stronger conditions at the end of the theorem's statement, for any fixed $a = (a_1,\dots,a_d) \in \mathbb{R}^d$,

$$\mathbb{P}\left(\frac{\widetilde{\Pi}_i^{(n)}((-\infty,a])}{\tau_n} - \Lambda_i((-\infty,a]) \to 0\right) = 1$$

by the law of large numbers. This extends to all rational a's, then to all a by approximation. The smoothing error is again controlled by Lemma 4.4.2.

4.4.2 *Consistency of Warp Functions and Inverses*

We next discuss the consistency of the warp and registration function estimators. These are key elements in order to align the observed point patterns $\widetilde{\Pi}_i$. Recall that we have consistent estimators $\widehat{\Lambda}_i$ for Λ_i and $\widehat{\lambda}_n$ for λ. Then $T_i = \mathbf{t}_\lambda^{\Lambda_i}$ is estimated by $\mathbf{t}_{\widehat{\lambda}_n}^{\widehat{\Lambda}_i}$ and T_i^{-1} is estimated by $\mathbf{t}_{\widehat{\Lambda}_i}^{\widehat{\lambda}_n}$. We will make the following extra assumptions that lead to more transparent statements (otherwise one needs to replace K with the set of Lebesgue points of the supports of λ and Λ_i).

Assumptions 4 (Strictly Positive Measures) *In addition to Assumptions 3 suppose that:*

1. *λ has a positive density on K (in particular, $\operatorname{supp}\lambda = K$);*
2. *T is almost surely surjective on $U = \operatorname{int}K$ (thus a homeomorphism of U).*

As a consequence $\operatorname{supp}\Lambda = \operatorname{supp}(T\#\lambda) = \overline{T(\operatorname{supp}\lambda)} = K$ almost surely.

Theorem 4.4.3 (Consistency of Optimal Maps) *Let Assumptions 4 be satisfied in addition to the hypotheses of Theorem 4.4.1. Then for any i such that* $\sigma_i^{(n)} \stackrel{P}{\to} 0$ *and any compact set* $S \subseteq \text{int}K$,

$$\sup_{x\in S} \|\widehat{T_i^{-1}}(x) - T_i^{-1}(x)\| \stackrel{\text{P}}{\to} 0, \qquad \sup_{x\in S} \|\widehat{T_i}(x) - T_i(x)\| \stackrel{\text{P}}{\to} 0.$$

Almost sure convergence can be obtained under the same provisions made at the end of the statement of Theorem 4.4.1.

A few technical remarks are in order. First and foremost, it is not clear that the two suprema are measurable. Even though T_i and T_i^{-1} are random elements in $C_b(U,\mathbb{R}^d)$, their estimators are only defined in an L_2 sense. The proof of Theorem 4.4.3 is done ω-wise. That is, for any ω in the probability space such that Theorem 4.4.1 holds, the two suprema vanish as $n \to \infty$. In other words, the convergence holds in outer probability or outer almost surely.

Secondly, assuming positive smoothing, the random measures $\widehat{\Lambda}_i$ are smooth with densities bounded below on K, so $\widehat{T_i^{-1}}$ are defined on the whole of U (possibly as set-valued functions on a Λ_i-null set). But the only known regularity result for $\widehat{\lambda}_n$ is an upper bound on its density (Proposition 3.1.8), so it is unclear what is its support and consequently what is the domain of definition of $\widehat{T}_i$.

Lastly, when the smoothing parameter σ is zero, $\widehat{T}_i$ and $\widehat{T_i^{-1}}$ are not defined. Nevertheless, Theorem 4.4.3 still holds in the set-valued formulation of Proposition 1.7.11, of which it is a rather simple corollary:

Proof (Proof of Theorem 4.4.3). The proof amounts to setting the scene in order to apply Proposition 1.7.11 of stability of optimal maps. We define

$$\mu_n = \widehat{\Lambda}_i; \qquad \nu_n = \widehat{\lambda}_n; \qquad \mu = \Lambda_i; \qquad \nu = \lambda; \qquad u_n = \widehat{T}_i^{-1}; \qquad u = T_i^{-1},$$

and verify the conditions of the proposition. The weak convergence of μ_n to μ and ν_n to ν is the conclusion of Theorem 4.4.1; the finiteness is apparent because K is compact and the uniqueness follows from the assumed absolute continuity of Λ_i. Since in addition T_i^{-1} is uniquely defined on $U = \text{int}K$ which is an open convex set, the restrictions on Ω in Proposition 1.7.11 are redundant. Uniform convergence of $\widehat{T}_i$ to T_i is proven in the same way.

Corollary 4.4.4 (Consistency of Point Pattern Registration) *For any i such that* $\sigma_i^{(n)} \stackrel{P}{\to} 0$,

$$W_2\left(\frac{\widehat{\Pi_i^{(n)}}}{N_i^{(n)}}, \frac{\Pi_i^{(n)}}{N_i^{(n)}}\right) \stackrel{\text{P}}{\to} 0.$$

The division by the number of observed points ensures that the resulting measures are probability measures; the relevant information is contained in the point patterns themselves, and is invariant under this normalisation.

Proof. The law of large numbers entails that $N_i^{(n)}/\tau_n \to 1$, so in particular $N_i^{(n)}$ is almost surely not zero when n is large. Since $\widehat{\Pi_i^{(n)}} = (\widehat{T_i^{-1}} \circ T_i)\#\Pi_i^{(n)}$, we have the upper bound

$$W_2^2\left(\frac{\widehat{\Pi_i^{(n)}}}{N_i^{(n)}}, \frac{\Pi_i^{(n)}}{N_i^{(n)}}\right) \le \int_K \|\widehat{T_i^{-1}}(T_i(x)) - x\|^2 \,\mathrm{d}\frac{\Pi_i^{(n)}}{N_i^{(n)}}.$$

Fix a compact $\Omega \subseteq \mathrm{int}K$ and split the integral to Ω and its complement. Then

$$\int_{K\setminus\Omega} \|\widehat{T_i^{-1}}(T_i(x)) - x\|^2 \,\mathrm{d}\frac{\Pi_i^{(n)}}{N_i^{(n)}} \le d_K^2 \frac{\Pi_i^{(n)}(K\setminus\Omega)}{\tau_n}\frac{\tau_n}{N_i^{(n)}} \stackrel{\text{as}}{\to} d_K^2\lambda(K\setminus\Omega),$$

by the law of large numbers, where d_K is the diameter of K. By writing $\mathrm{int}K$ as a countable union of compact sets (and since λ is absolutely continuous), this can be made arbitrarily small by choice of Ω.

We can easily bound the integral on Ω by

$$\int_\Omega \|\widehat{T_i^{-1}}(T_i(x)) - x\|^2 \,\mathrm{d}\frac{\Pi_i^{(n)}}{N_i^{(n)}} \le \sup_{x\in\Omega} \|\widehat{T_i^{-1}}(T_i(x)) - x\|^2 = \sup_{y\in T_i(\Omega)} \|\widehat{T_i^{-1}}(y) - T_i^{-1}(y)\|^2.$$

But $T_i(\Omega)$ is a compact subset of $U = \mathrm{int}K$, because $T_i \in C_b(U,U)$. The right-hand side therefore vanishes as $n \to \infty$ by Theorem 4.4.3, and this completes the proof.

Possible extensions pertaining to the boundary of K are discussed on page 33 of the supplement.

4.5 Illustrative Examples

In this section, we illustrate the estimation framework put forth in this chapter by considering an example of a structural mean λ with a bimodal density on the real line. The unwarped point patterns Π originate from Poisson processes with mean measure λ and, consequently, the warped points $\widetilde{\Pi}$ are Cox processes (see Sect. 4.1.2). Another scenario involving triangular densities can be found in Panaretos and Zemel [100].

4.5.1 Explicit Classes of Warp Maps

As a first step, we introduce a class of random warp maps satisfying Assumptions 2, that is, increasing maps that have as mean the identity function. The construction is a mixture version of similar maps considered by Wang and Gasser [128, 129].

For any integer k define $\zeta_k : [0,1] \to [0,1]$ by

$$\zeta_0(x) = x, \qquad \zeta_k(x) = x - \frac{\sin(\pi k x)}{|k|\pi}, \quad k \in \mathbb{Z}\setminus\{0\}. \tag{4.4}$$

Clearly $\zeta_k(0) = 0$, $\zeta_k(1) = 1$ and ζ_k is smooth and strictly increasing for all k. Figure 4.4a plots ζ_k for $k = -3, \ldots, 3$. To make ζ_k a random function, we let k be an integer-valued random variable. If the latter is symmetric, then we have

$$\mathbb{E}\left[\zeta_k(x)\right] = x, \qquad x \in [0,1].$$

By means of mixtures, we replace this discrete family by a continuous one: let $J > 1$ be an integer and $V = (V_1, \ldots, V_J)$ be a random vector following the flat Dirichlet distribution (uniform on the set of nonnegative vectors with $v_1 + \cdots + v_J = 1$). Take independently k_j following the same distribution as k and define

$$T(x) = \sum_{j=1}^{J} V_j \zeta_{k_j}(x). \tag{4.5}$$

Since V_j is positive, T is increasing and as (V_j) sums up to unity T has mean identity. Realisations of these warp functions are given in Fig. 4.4b and c for $J = 2$ and $J = 10$, respectively. The parameters (k_j) were chosen as symmetrised Poisson random variables: each k_j has the law of XY with X Poisson with mean 3 and $\mathbb{P}(Y = 1) = \mathbb{P}(Y = -1) = 1/2$ for Y and X independent. When $J = 10$ is large, the function T deviates only mildly from the identity, since a law of large numbers begins to take effect. In contrast, $J = 2$ yields functions that are quite different from the identity. Thus, it can be said that the parameter J controls the variance of the random warp function T.

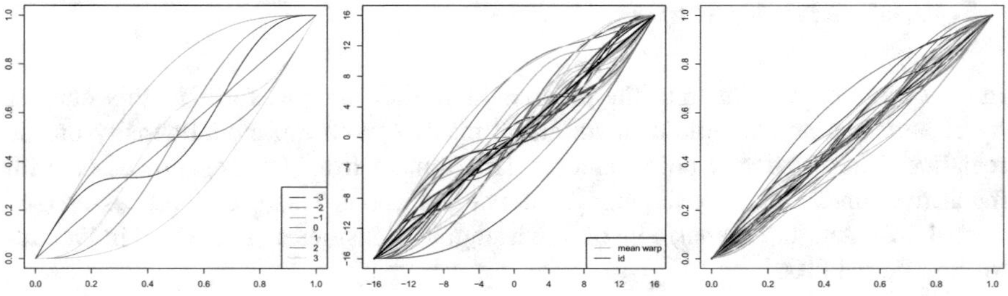

(a) (b) (c)

Fig. 4.4: (**a**) The functions $\{\zeta_{-3}, \ldots, \zeta_3\}$; (**b**) realisations of T defined by (4.5) with $J = 2$ and k_j symmetrisations of Poisson random variables with mean 3; (**c**) realisations of T defined by (4.5) with $J = 10$ and k_j as in (**b**)

4.5.2 Bimodal Cox Processes

Let the structural mean measure λ be a mixture of a bimodal Gaussian distribution (restricted to $K = [-16,16]$) and a beta background on the interval $[-12,12]$, so that mass is added at the centre of K but not near the boundary. In symbols this is given as follows. Let φ be the standard Gaussian density and let $\beta_{\alpha,\beta}$ denote the density of a the beta distribution with parameters α and β. Then λ is chosen as the measure with density

$$f(x) = \frac{1-\varepsilon}{2}[\varphi(x-8)+\varphi(x+8)] + \frac{\varepsilon}{24}\beta_{1.5,1.5}\left(\frac{x+12}{24}\right), \qquad x \in [-16,16], \tag{4.6}$$

where $\varepsilon \in [0,1]$ is the weight of the beta background. (We ignore the loss of a negligible amount of mass due to the restriction of the Gaussians to $[-16,16]$.) Plots of the density and distribution functions are given in Fig. 4.5.

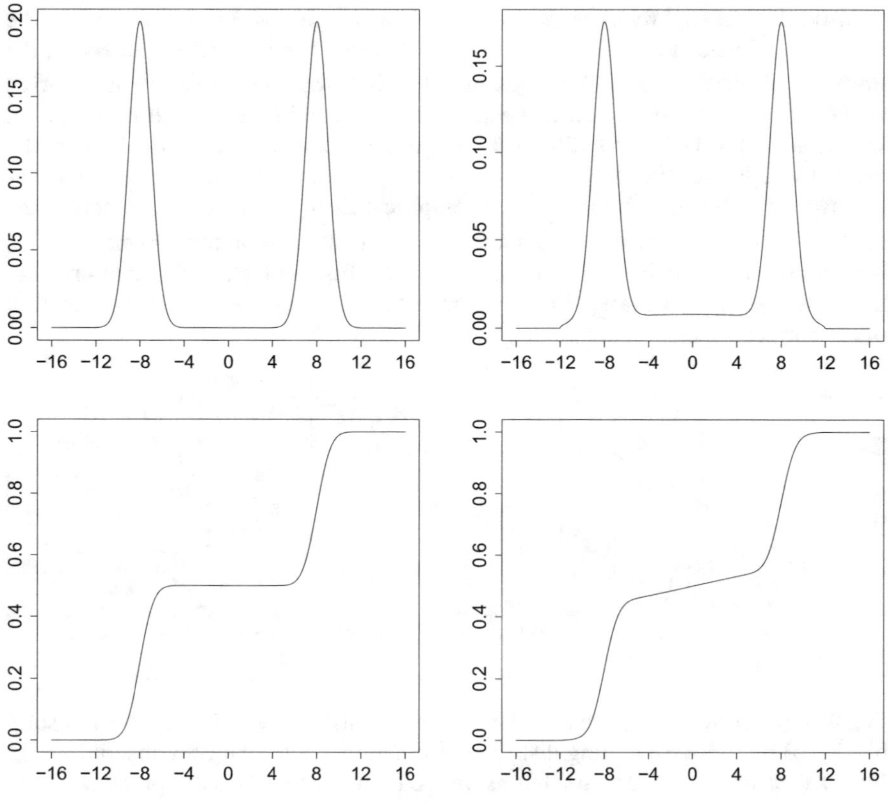

Fig. 4.5: Density and distribution functions corresponding to (4.6) with $\varepsilon = 0$ and $\varepsilon = 0.15$

The main criterion for the quality of our regularised Fréchet–Wasserstein estimator will be its success in discerning the two modes at ± 8; these will be smeared by the phase variation arising from the warp functions.

We next simulated 30 independent Poisson processes with mean measure λ, $\varepsilon = 0.1$, and total intensity (expected number of points) $\tau = 93$. In addition, we generated warp functions as in (4.5) but rescaled to $[-16, 16]$; that is, having the same law as the functions

$$x \mapsto 32T\left(\frac{x+16}{32}\right) - 16$$

from K to K. These cause rather violent phase variation, as can be seen by the plots of the densities and distribution functions of the conditional measures $\Lambda = T\#\lambda$ presented in Fig. 4.6a and b; the warped points themselves are displayed in Fig. 4.6c.

Using these warped point patterns, we construct the regularised Fréchet–Wasserstein estimator employing the procedure described in Sect. 4.3. Each $\widetilde{\Pi}_i$ was smoothed with a Gaussian kernel and bandwidth chosen by unbiased cross validation. We deviate slightly from the recipe presented in Sect. 4.3 by not restricting the resulting estimates to the interval $[-16, 16]$, but this has no essential effect on the finite sample performance. The regularised Fréchet–Wasserstein estimator $\widehat{\lambda}_n$ serves as the estimator of the structural mean λ and is shown in Fig. 4.7a. It is contrasted with λ at the level of distribution functions, as well as with the empirical arithmetic mean; the latter, the *naive estimator*, is calculated by ignoring the warping and simply averaging linearly the (smoothed) empirical distribution functions across the observations. We notice that $\widehat{\lambda}_n$ is rather successful at locating the two modes of λ, in contrast with the naive estimator that is more diffuse. In fact, its distribution function increases approximately linearly, suggesting a nearly constant density instead of the correct bimodal one.

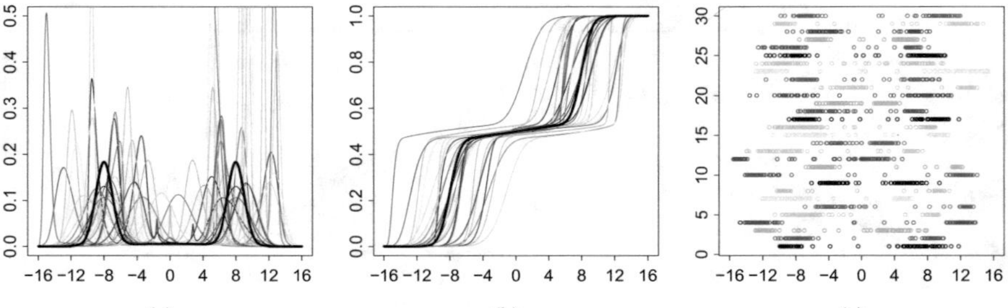

Fig. 4.6: (**a**) 30 warped bimodal densities, with density of λ given by (4.6) in solid black; (**b**) their corresponding distribution functions, with that of λ in solid black; (**c**) 30 Cox processes, constructed as warped versions of Poisson processes with mean intensity $93f$ using as warp functions the rescaling to $[-16.16]$ of (4.5)

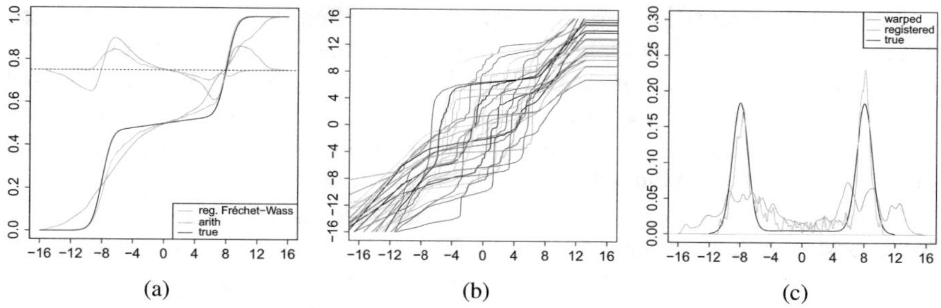

Fig. 4.7: (**a**) Comparison between the regularised Fréchet–Wasserstein estimator, the empirical arithmetic mean, and the true distribution function, including residual curves centred at $y = 3/4$; (**b**) The estimated warp functions; (**c**) Kernel estimates of the density function f of the structural mean, based on the warped and registered point patterns

Estimators of the warp maps $\widehat{T_i}$, depicted in Fig. 4.7b, and their inverses, are defined as the optimal maps between $\widehat{\lambda}_n$ and the estimated conditional mean measures, as explained in Sect. 4.3.4. Then we register the point patterns by applying to them the inverse estimators $\widehat{T_i^{-1}}$ (Fig. 4.8). Figure 4.7c gives two kernel estimators of the density of λ constructed from a superposition of all the warped points and all the registered ones. Notice that the estimator that uses the registered points is much more successful than the one using the warped ones in discerning the two density peaks. This is not surprising after a brief look at Fig. 4.8, where the unwarped, warped, and registered points are displayed. Indeed, there is very high concentration of registered points around the true location of the peaks, ± 8. This is not the case for the warped points because of the phase variation that translates the centres of concentration for each individual observation. It is important to remark that the fluctuations in the density estimator in Fig. 4.7c are not related to the registration procedure, and could be reduced by a better choice of bandwidth. Indeed, our procedure does not attempt to estimate the density, but rather the distribution function.

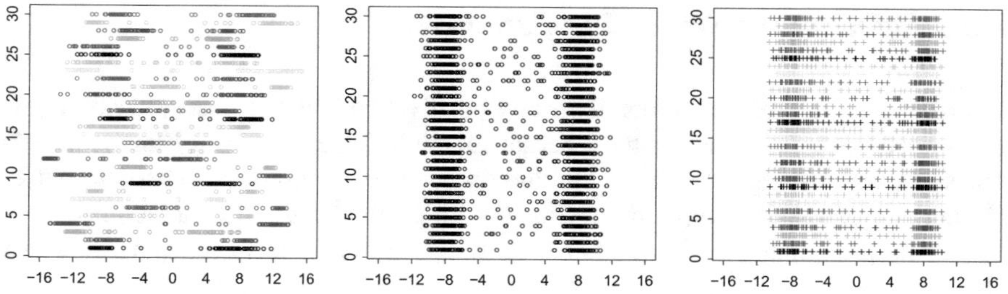

Fig. 4.8: Bimodal Cox processes: (**a**) the observed warped point processes; (**b**) the unobserved original point processes; (**c**) the registered point processes

Figure 4.9 presents a superposition of the regularised Fréchet–Wasserstein estimators for 20 independent replications of the experiment, contrasted with a similar superposition for the naive estimator. The latter is clearly seen to be biased around the two peaks, while the regularised Fréchet–Wasserstein seems approximately unbiased, despite presenting fluctuations. It always captures the bimodal nature of the density, as is seen from the two clear elbows in each realisation.

To illustrate the consistency of the regularised Fréchet–Wasserstein estimator $\widehat{\lambda}_n$ for λ as shown in Theorem 4.4.1, we let the number of processes n as well as the expected number of observed point per process τ to vary. Figures 4.10 and 4.11 show the sampling variation of $\widehat{\lambda}_n$ for different values of n and τ. We observe that as either of these increases, the realisations $\widehat{\lambda}_n$ indeed approach λ. The figures suggest that, in this scenario, the amplitude variation plays a stronger role than the phase variation, as the effect of τ is more substantial.

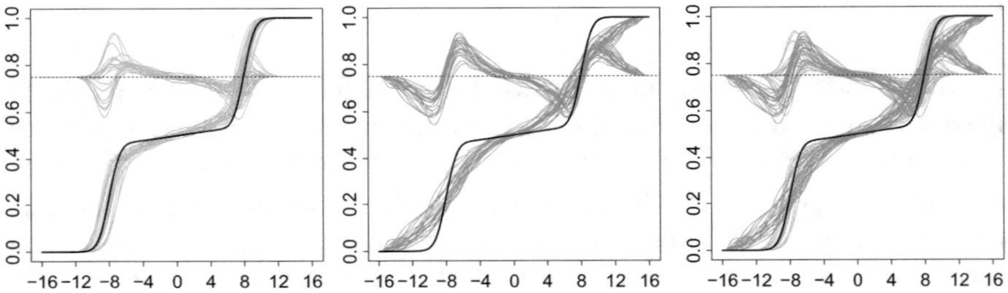

Fig. 4.9: (**a**) Sampling variation of the regularised Fréchet–Wasserstein mean $\widehat{\lambda}_n$ and the true mean measure λ for 20 independent replications of the experiment; (**b**) sampling variation of the arithmetic mean, and the true mean measure λ for the same 20 replications; (**c**) superposition of (**a**) and (**b**). For ease of comparison, all three panels include residual curves centred at $y = 3/4$

4.5.3 Effect of the Smoothing Parameter

In order to work with measures of strictly positive density, the observed point patterns have been smoothed using a kernel function. This necessarily incurs an additional bias that depends on the bandwidth σ_i. The asymptotics (Theorem 4.4.1) guarantee the consistency of the estimators, in particular the regularised Fréchet–Wasserstein estimator $\widehat{\lambda}_n$, provided that $\max_{i=1}^n \sigma_i \to 0$. In our simulations, we choose σ_i in a data-driven way by employing unbiased cross validation. To gauge for the effect of the smoothing, we carry out the same estimation procedure but with σ_i multiplied by a parameter s. Figure 4.12 presents the distribution function of $\widehat{\lambda}_n$ as a function of s. Interestingly, the curves are nearly identical as long as $s \leq 1$, whereas when $s > 1$, the bias becomes more substantial.

These findings are reaffirmed in Fig. 4.13 that show the registered point processes again as a function of s. We see that only minor differences are present as s varies from 0.1 to 1, for example, in the grey (8), black (17), and green (19) processes. When $s = 3$, the distortion becomes quite more substantial. This phenomenon repeats itself across all combinations of n, τ, and s tested.

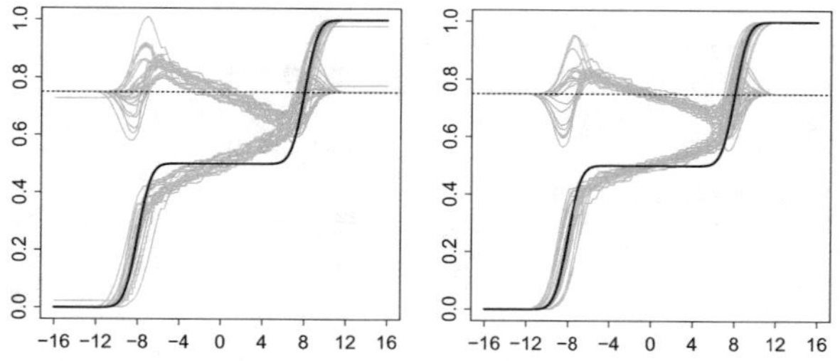

Fig. 4.10: Sampling variation of the regularised Fréchet–Wasserstein mean $\widehat{\lambda}_n$ and the true mean measure λ for 20 independent replications of the experiment, with $\varepsilon = 0$ and $n = 30$. Left: $\tau = 43$; middle: $\tau = 93$; right: $\tau = 143$. For ease of comparison, all three panels include residual curves centred at $y = 3/4$

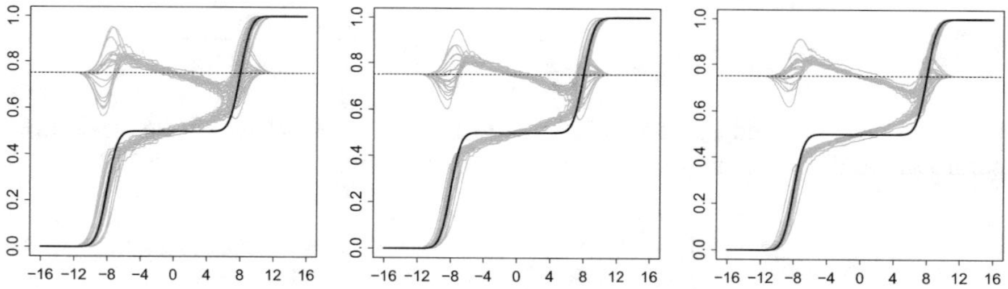

Fig. 4.11: Sampling variation of the regularised Fréchet–Wasserstein mean $\widehat{\lambda}_n$ and the true mean measure λ for 20 independent replications of the experiment, with $\varepsilon = 0$ and $\tau = 93$. Left: $n = 30$; middle: $n = 50$; right: $n = 70$. For ease of comparison, all three panels include residual curves centred at $y = 3/4$.

4.6 Convergence Rates and a Central Limit Theorem on the Real Line

Since the conditional mean measures Λ_i are discretely observed, the rate of convergence of our estimators will be affected by the rate at which the number of observed points per process $N_i^{(n)}$ increases to infinity. The latter is controlled by the next lemma, which is valid for any complete separable metric space $\mathscr{X}$.

Lemma 4.6.1 (Number of Points Grows Linearly) *Let* $N_i^{(n)} = \Pi_i^{(n)}(\mathscr{X})$ *denote the total number of observed points. If* $\tau_n / \log n \to \infty$*, then there exists a constant* $C_\Pi > 0$*, depending only on the distribution of* Π*, such that almost surely*

$$\liminf_{n\to\infty} \frac{\min_{1\le i\le n} N_i^{(n)}}{\tau_n} \ge C_\Pi .$$

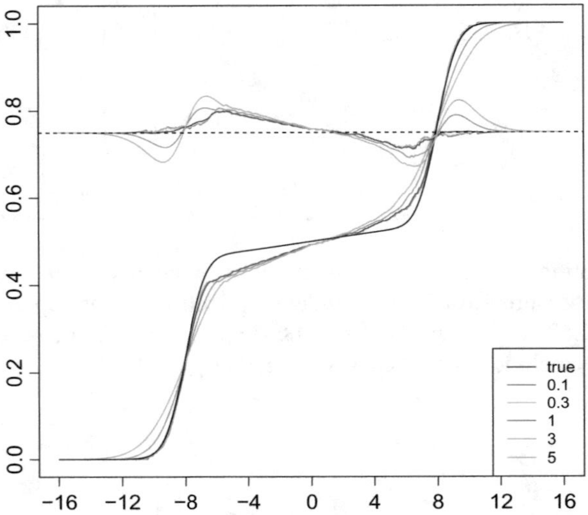

Fig. 4.12: Regularised Fréchet–Wasserstein mean as a function of the smoothing parameter multiplier s, including residual curves. Here, $n = 30$ and $\tau = 143$

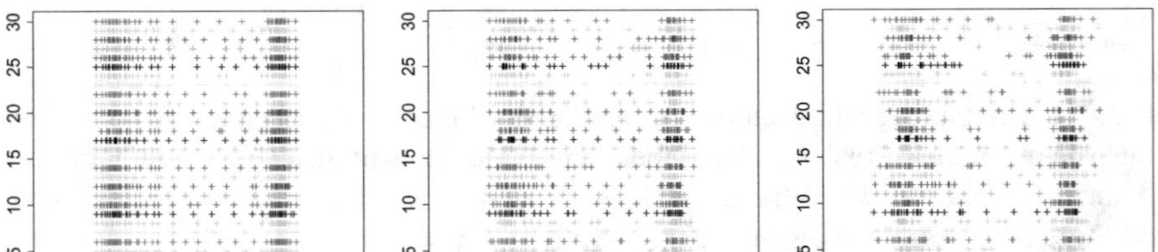

Fig. 4.13: Registered point processes as a function of the smoothing parameter multiplier s. Left: $s = 0.1$; middle: $s = 1$; right: $s = 3$. Here, $n = 30$ and $\tau = 43$

In particular, there are no empty point processes, so the normalisation is well-defined. If Π *is a Poisson process, then we have the more precise result*

$$\lim_{n\to\infty} \frac{\min_{1\le i\le n} N_i^{(n)}}{\tau_n} = 1 \qquad \textit{almost surely.}$$

Remark 4.6.2 *One can also show that the limit superior of the same quantity is bounded by a constant C'_Π. If $\tau_n/\log n$ is bounded below, then the same result holds but with worse constants. If only $\tau_n \to \infty$, then the result holds for each i separately but in probability.*

The proof is a simple application of Chernoff bounds; see page 108 in the supplement.

With Lemma 4.6.1 under our belt, we can replace terms of the order $\min_i N_i^{(n)}$ by the more transparent order τ_n. As in the consistency proof, the idea is to write

$$F - \widehat{F}_n = (F - F_n) + (F_n - \tilde{F}_n) + (\tilde{F}_n - \widehat{F}_n)$$

and control each term separately. The first term corresponds to the phase variation, and comes from the approximation of the theoretical expectation F by a sample mean F_n. The second term is associated with the amplitude variation resulting from observing Λ_i discretely. The third term can be viewed as the bias incurred by the smoothing procedure. Accordingly, the rate at which $\widehat{\lambda}_n$ converges to λ is a sum of three separate terms. We recall the standard $O_{\mathbb{P}}$ terminology: if X_n and Y_n are random variables, then $X_n = O_{\mathbb{P}}(Y_n)$ means that the sequence (X_n/Y_n) is *bounded in probability*, which by definition is the condition

$$\forall \varepsilon > 0\ \exists M: \quad \sup_n \mathbb{P}\left(\left|\frac{X_n}{Y_n}\right| > M\right) < \varepsilon.$$

Instead of $X_n = O_{\mathbb{P}}(Y_n)$, we will sometimes write $Y_n \geq O_{\mathbb{P}}(X_n)$. The former notation emphasises the condition that X_n grows no faster than Y_n, while the latter stresses that Y_n grows at least as fast as X_n (which is of course the same assertion). Finally, $X_n = o_{\mathbb{P}}(Y_n)$ means that $X_n/Y_n \to 0$ in probability.

Theorem 4.6.3 (Convergence Rates on $\mathbb{R}$) *Suppose in addition to Assumptions 3 that $d = 1$, $\tau_n/\log n \to \infty$ and that Π is either a Poisson process or a binomial process. Then*

$$W_2(\widehat{\lambda}_n, \lambda) \leq O_{\mathbb{P}}\left(\frac{1}{\sqrt{n}}\right) + O_{\mathbb{P}}\left(\frac{1}{\sqrt[4]{\tau_n}}\right) + O_{\mathbb{P}}(\sigma_n), \qquad \sigma_n = \frac{1}{n}\sum_{i=1}^n \sigma_i^{(n)},$$

where all the constants in the $O_{\mathbb{P}}$ terms are explicit.

Remark 4.6.4 *Unlike classical density estimation, no assumptions on the rate of decay of σ_n are required, because we only need to estimate the distribution function and not the derivative. If the smoothing parameter is chosen to be $\sigma_i^{(n)} = [N_i^{(n)}]^{-\alpha}$ for some $\alpha > 0$ and $\tau_n/\log n \to \infty$, then by Lemma 4.6.1 $\sigma_n \leq \max_{1\leq i\leq n}\sigma_i^{(n)} = O_{\mathbb{P}}(\tau_n^{-\alpha})$. For example, if* Rosenblatt's rule *$\alpha = 1/5$ is employed, then the $O_{\mathbb{P}}(\sigma_n)$ term can be replaced by $O_{\mathbb{P}}(1/\sqrt[5]{\tau_n})$.*

One can think about the parameter τ as separating the *sparse* and *dense* regimes as in classical functional data analysis (see also Wu et al. [132]). If τ is bounded, then the setting is *ultra sparse* and consistency cannot be achieved. A sparse regime can be defined as the case where $\tau_n \to \infty$ but slower than $\log n$. In that case, consistency is guaranteed, but some point patterns will be empty. The *dense* regime can be defined as $\tau_n \gg n^2$, in which case the amplitude variation is negligible asymptotically when compared with the phase variation.

The exponent $-1/4$ of τ_n can be shown to be optimal without further assumptions, but it can be improved to $-1/2$ if $\mathbb{P}(f_\Lambda \geq \varepsilon \text{ on } K) = 1$ for some $\varepsilon > 0$, where f_Λ is the density of Λ (see Sect. 4.7). In terms of T, the condition is that $\mathbb{P}(T' \geq \varepsilon) = 1$ for some ε and λ has a density bounded below. When this is the case, τ_n needs to compared with n rather than n^2 in the next paragraph and the next theorem.

Theorem 4.6.3 provides conditions for the optimal parametric rate $\sqrt{n}$ to be achieved: this happens if we set σ_n to be of the order $O_{\mathbb{P}}(n^{-1/2})$ or less and if τ_n is of the order n^2 or more. But if the last two terms in Theorem 4.6.3 are *negligible* with respect to $n^{-1/2}$, then a sort of *central limit theorem* holds for $\widehat{\lambda}_n$:

Theorem 4.6.5 (Asymptotic Normality) *In addition to the conditions of Theorem 4.6.3, assume that $\tau_n/n^2 \to \infty$, $\sigma_n = o_{\mathbb{P}}(n^{-1/2})$ and λ possesses an invertible distribution function F_λ on K. Then*

$$\sqrt{n}\left(\mathbf{t}_{\lambda}^{\widehat{\lambda}_n} - \mathbf{i}\right) \longrightarrow Z \quad \textit{weakly in } L_2(\lambda),$$

for a zero-mean Gaussian process Z with the same covariance operator of T (the latter viewed as a random element in $L_2(\lambda)$), namely with covariance kernel

$$\kappa(x,y) = \operatorname{cov}\Big\{T(x), T(y)\Big\}.$$

If the density f_λ exists and is (piecewise) continuous and bounded below on K, then the weak convergence also holds in $L_2(K)$.

In view of Sect. 2.3, Theorem 4.6.5 can be interpreted as asymptotic normality of $\widehat{\lambda}_n$ in the *tangential* sense: $\sqrt{n}\log_\lambda(\widehat{\lambda}_n)$ converges to a Gaussian random element in the tangent space Tan_λ, which is a subset of $L_2(\lambda)$. The additional smoothness conditions allow to switch to the space $L_2(K)$, which is independent of the unknown template measure λ.

See pages 109 and 110 in the supplement for detailed proofs of these theorems. Below we sketch the main ideas only.

Proof (Proof of Theorem 4.6.3). The quantile formula $W_2(\gamma, \theta) = \|F_\theta^{-1} - F_\gamma^{-1}\|_{L^2(0,1)}$ from Sect. 1.5 and the average quantile formula for the Fréchet mean (Sect. 3.1.4) show that the oracle empirical mean $F_{\overline{\lambda}_n}^{-1}$ follows a central limit theorem in $L_2(0,1)$. Since we work in the Hilbert space $L_2(0,1)$, Fréchet means are simple averages, so the errors in the Fréchet mean have the same rate as the errors in the Fréchet functionals. The smoothing term is easily controlled by Lemma 4.4.2.

Controlling the amplitude term is more difficult. Bounds can be given using the machinery sketched in Sect. 4.7, but we give a more elementary proof by reducing to the 1-Wasserstein case (using (2.2)), which can be more easily handled in terms of distributions functions (Corollary 1.5.3).

Proof (Proof of Theorem 4.6.5). The hypotheses guarantee that the amplitude and smoothing errors are negligible and

$$\sqrt{n}\left(F_{\widehat{\lambda}_n}^{-1} - F_{\lambda}^{-1}\right) \to GP \quad \text{weakly in } L_2(0,1),$$

where GP is the Gaussian process defined in the proof of Theorem 4.6.3. One then employs a composition with F_λ.

4.7 Convergence of the Empirical Measure and Optimality

One may find the term $O_{\mathbb{P}}(1/\sqrt[4]{\tau_n})$ in Theorem 4.6.3 to be somewhat surprising, and expect that it ought to be $O_{\mathbb{P}}(1/\sqrt{\tau_n})$. The goal of this section is to show why the rate $1/\sqrt[4]{\tau_n}$ is optimal without further assumptions and discuss conditions under which it can be improved to the optimal rate $1/\sqrt{\tau_n}$. For simplicity, we concentrate on the case $\tau_n = n$ and assume that the point process Π is binomial; the Poisson case being easily obtained from the simplified one (using Lemma 4.6.1). We are thus led to study rates of convergence of empirical measures in the Wasserstein space. That is to say, for a fixed exponent $p \geq 1$ and a fixed measure $\mu \in \mathcal{W}_p(\mathcal{X})$, we consider independent random variables $X_1, \dots$ with law μ and the *empirical measure* $\mu_n = n^{-1}\sum_{i=1}^n \delta\{X_i\}$. The first observation is that $\mathbb{E}W_p(\mu, \mu_n) \to 0$:

Lemma 4.7.1 *Let $\mu \in P(\mathcal{X})$ be any measure. Then*

$$\mathbb{E}W_p(\mu,\mu_n)\begin{cases} = \infty & \mu \notin \mathcal{W}_p(\mathcal{X}) \\ \to 0 & \mu \in \mathcal{W}_p(\mathcal{X}). \end{cases}$$

Proof. This result has been established in an almost sure sense in Proposition 2.2.6. To extend to convergence in expectation observe that

$$W_p^p(\mu,\mu_n) \leq \int_{\mathcal{X}^2} \|x-y\|^p \, \mathrm{d}\mu \otimes \mu_n(x,y) = \frac{1}{n}\sum_{i=1}^n \int_{\mathcal{X}} \|x - X_i\|^p \, \mathrm{d}\mu(x).$$

Thus, the random variable $0 \leq Y_n = W_p^p(\mu,\mu_n)$ is bounded by the sample average Z_n of a random variable $V = \int_{\mathcal{X}} \|x - X_1\|^p \, \mathrm{d}\mu(x)$ that has a finite expectation. A version of the dominated converge theorem (given on page 111 in the supplement) implies that $\mathbb{E}Y_n \to 0$. Now invoke Jensen's inequality.

Remark 4.7.2 *The sequence $\mathbb{E}W_p(\mu,\mu_n)$ is not monotone, as the simple example $\mu = (\delta_0 + \delta_1)/2$ shows (see page 111 in the supplement).*

The next question is how quickly $\mathbb{E}W_p(\mu,\mu_n)$ vanishes when $\mu \in \mathcal{W}_p(\mathcal{X})$. We shall begin with two simple general lower bounds, then discuss upper bounds in the one-dimensional case, put them in the context of Theorem 4.6.3, and finally briefly touch the d-dimensional case.

Lemma 4.7.3 ($\sqrt{n}$ Lower Bound) *Let $\mu \in P(\mathscr{X})$ be nondegenerate. Then there exists a constant $c(\mu) > 0$ such that for all $p \geq 1$ and all n*

$$\mathbb{E}W_p(\mu_n, \mu) \geq \frac{c(\mu)}{\sqrt{n}}.$$

Proof. Let $X \sim \mu$ and let $a \neq b$ be two points in the support μ. Consider $f(x) = \min(1, \|x-a\|)$, a bounded 1-Lipschitz function such that $f(a) = 0 < f(b)$. Then

$$\sqrt{n}\mathbb{E}W_p(\mu_n, \mu) \geq \sqrt{n}\mathbb{E}W_1(\mu_n, \mu) \geq \mathbb{E}\left|n^{-1/2}\sum_{i=1}^{n} f(X_i) - \mathbb{E}f(X)\right| \to \sqrt{\frac{2\mathrm{var} f(X)}{\pi}} > 0$$

by the central limit theorem and the Kantorovich–Rubinstein theorem (1.11).

For discrete measures, the rates scale badly with p. More generally:

Lemma 4.7.4 (Separated Support) *Suppose that there exist Borel sets $A, B \subset \mathscr{X}$ such that $\mu(A \cup B) = 1$,*

$$\mu(A)\mu(B) > 0 \qquad \textit{and} \qquad d_{\min} = \inf_{x\in A, y\in B} \|x-y\| > 0.$$

Then for any $p \geq 1$ there exists $c_p(\mu) > 0$ such that $\mathbb{E}W_p(\mu_n, \mu) \geq c_p(\mu)n^{-1/(2p)}$.

Any nondegenerate finitely discrete measure μ satisfies this condition, and so do "non-pathological" countably discrete ones. (An example of a "pathological" measure is one assigning positive mass to any rational number.)

Proof. Let $k \sim B(n, q = \mu(A))$ denote the number of points from the sample $(X_1, \ldots, X_n)$ that fall in A. Then a mass of $|k/n - q|$ must travel between A and B, a distance of at least $d_{\min}$. Thus, $W_p^p(\mu_n, \mu) \geq d_{\min}^p |k/n - q|$, and the result follows from the central limit theorem for k; see page 112 in the supplement for the full details.

These lower bounds are valid on any separable metric space. On the real line, it is easy to obtain a sufficient condition for the optimal rate $n^{-1/2}$ to be attained for W_1: since $F_n(t) \sim B(n, F(t))$ has variance $F(t)(1-F(t))/n$, we have (by Fubini's theorem and Jensen's inequality)

$$\mathbb{E}W_1(\mu_n, \mu) = \int_{\mathbb{R}} \mathbb{E}|F_n(t) - F(t)|\,\mathrm{d}t \leq n^{-1/2}\int_{\mathbb{R}} \sqrt{F(t)(1-F(t))}\,\mathrm{d}t,$$

so that $W_1(\mu_n, \mu)$ is of the optimal order $n^{-1/2}$ if

$$J_1(\mu) := \int_{\mathbb{R}} \sqrt{F(t)(1-F(t))}\,\mathrm{d}t < \infty.$$

Since the integrand is bounded by $1/2$, this is certainly satisfied if μ is compactly supported. The J_1 condition is essentially a moment condition, since for any $\delta > 0$, we have for $X \sim \mu$ that $\mathbb{E}|X|^{2+\delta} < \infty \Longrightarrow J_1(\mu) < \infty \Longrightarrow \mathbb{E}|X|^2 < \infty$. It turns out that this condition is necessary, and has a more subtle counterpart for any $p \geq 1$. Let f denote the density of the absolutely continuous part of μ (so $f \equiv 0$ if μ is discrete).

Theorem 4.7.5 (Rate of Convergence of Empirical Measures) *Let $p \geq 1$ and $\mu \in \mathscr{W}_p(\mathbb{R})$. The condition*

$$J_p(\mu) = \int_{\mathbb{R}} \frac{[F(t)(1-F(t))]^{p/2}}{[f(t)]^{p-1}} \, \mathrm{d}t < \infty, \qquad (0^0 = 1)$$

is necessary and sufficient for $\mathbb{E}W_p(\mu_n, \mu) = O(n^{-1/2})$.

See Bobkov and Ledoux [25, Theorem 5.10] for a proof for the J_p condition, and Theorems 5.1 and 5.3 for the values of the constants and a stronger result.

When $p > 1$, for $J_p(\mu)$ to be finite, the support of μ must be connected; this is not needed when $p = 1$. Moreover, the J_p condition is satisfied when f is bounded below (in which case the support of μ must be compact). However, smoothness alone does not suffice, even for measures with positive density on a compact support. More precisely, we have:

Proposition 4.7.6 *For any rate $\varepsilon_n \to 0$ there exists a measure μ on $[-1,1]$ with positive C^∞ density there, and such that for all n*

$$\mathbb{E}W_p(\mu_n, \mu) \geq C(p, \mu) n^{-1/(2p)} \varepsilon_n.$$

The rate $n^{-1/(2p)}$ from Lemma 4.7.4 is the worst among compactly supported measures on $\mathbb{R}$. Indeed, by Jensen's inequality and (2.2), for any $\mu \in P([0,1])$,

$$\mathbb{E}W_p(\mu_n, \mu) \leq \left[\mathbb{E}W_p^p(\mu_n, \mu)\right]^{1/p} \leq [\mathbb{E}W_1(\mu_n, \mu)]^{1/p} \leq n^{-1/(2p)}.$$

The proof of Proposition 4.7.6 is done by "smoothing" the construction in Lemma 4.7.4, and is given on page 113 in the supplement.

Let us now put this in the context of Theorem 4.6.3. In the binomial case, since each $\Pi_i^{(n)}$ and each Λ_i are independent, we have

$$\mathbb{E}W_2(\Lambda_i, \widetilde{\Lambda}_i)|\Lambda_i \leq \sqrt{2J_2(\Lambda_i)} \frac{1}{\sqrt{\tau_n}}.$$

(In the Poisson case, we need to condition on $N_i^{(n)}$ and then estimate its inverse square root as is done in the proof of Theorem 4.6.3.) Therefore, a sufficient condition for the rate $1/\sqrt{\tau_n}$ to hold is that $\mathbb{E}\sqrt{J_2(\Lambda)} < \infty$ and a necessary condition is that $\mathbb{P}(\sqrt{J_2(\Lambda)} < \infty) = 1$. These hold if there exists $\delta > 0$ such that with probability one Λ has a density bounded below by δ. Since $\Lambda = T\#\lambda$, this will happen provided that λ itself has a bounded below density and T has a bounded below derivative. Bigot et al. [23] show that the rate $\sqrt{\tau_n}$ cannot be improved.

We conclude by proving a lower bound for absolutely continuous measures and stating, without proof, an upper bound.

Proposition 4.7.7 *Let $\mu \in \mathscr{W}_1(\mathbb{R}^d)$ have an absolutely continuous part with respect to Lebesgue measure, and let ν_n be any discrete measure supported on n points (or less). Then there exists a constant $C(\mu) > 0$ such that*

$$W_p(\mu, \nu_n) \geq W_1(\mu, \nu_n) \geq C(\mu) n^{-1/d}.$$

Proof. Let f be the density of the absolutely continuous part μ_c, and observe that for some finite number M,

$$2\delta = \mu_c(\{x : f(x) \leq M\}) > 0.$$

Let $x_1, \dots, x_n$ be the support points of ν_n and $\varepsilon > 0$. Let $\mu_{c,M}$ be the restriction of μ_c to the set where the density is smaller than M. The union of balls $B_\varepsilon(x_i)$ has $\mu_{c,M}$-measure of at most

$$M \sum_{i=1}^{n} \mathrm{Leb}(B_\varepsilon(x_i)) = Mn\varepsilon^d \mathrm{Leb}_d(B_1(0)) = Mn\varepsilon^d C_d = \delta,$$

if $\varepsilon^d = \delta(nMC_d)^{-1}$. Thus, a mass $2\delta - \delta = \delta$ must travel more than ε from ν_n to μ in order to cover $\mu_{c,M}$. Hence

$$W_1(\nu_n, \mu) \geq \delta\varepsilon = \delta(\delta/MC_d)^{1/d}\, n^{-1/d}.$$

The lower bound holds because we need ε^{-d} balls of radius ε in order to cover a sufficiently large fraction of the mass of μ. The determining quantity for *upper* bounds on the empirical Wasserstein distance is the *covering numbers*

$$N(\mu, \varepsilon, \tau) = \text{ minimal number of balls whose union has } \mu \text{ mass } \geq 1 - \tau.$$

Since μ is tight, these are finite for all $\varepsilon, \tau > 0$, and they increase as ε and τ approach zero. To put the following bound in context, notice that if μ is compactly supported on $\mathbb{R}^d$, then $N(\mu, \varepsilon, 0) \leq K\varepsilon^{-d}$.

Theorem 4.7.8 *If for some $d > 2p$, $N(\mu, \varepsilon, \varepsilon^{dp/(d-2p)}) \leq K\varepsilon^{-d}$, then $\mathbb{E}W_p \leq C_p n^{-1/d}$.*

Comparing this with the lower bound in Lemma 4.7.4, we see that in the high-dimensional regime $d > 2p$, absolutely continuous measures have a worse rate than discrete ones. In the low-dimensional regime $d < 2p$, the situation is opposite. We also obtain that for $d > 2$ and a compactly supported absolutely continuous $\mu \in \mathscr{W}_1(\mathbb{R}^d)$, $\mathbb{E}W_1(\mu_n, \mu) \sim n^{-1/d}$.

4.8 Bibliographical Notes

Our exposition in this chapter closely follows the papers Panaretos and Zemel [100] and Zemel and Panaretos [134].

Books on functional data analysis include Ramsay and Silverman [109, 110], Ferraty and Vieu [51], Horváth and Kokoszka [70], and Hsing and Eubank [71], and a recent review is also available (Wang et al. [127]). The specific topic of amplitude and phase variation is discussed in [110, Chapter 7] and [127, Section 5.2]. The next paragraph gives some selective references.

One of the first functional registration techniques employed dynamic programming (Wang and Gasser [128]) and dates back to Sakoe and Chiba [118]. Landmark registration consists of identifying salient features for each curve, called *landmarks*, and aligning them (Gasser and Kneip [61]; Gervini and Gasser [63]). In pairwise synchronisation (Tang and Müller [122]) one aligns each pair of curves and then derives an estimator of the warp functions by linear averaging of the pairwise registration maps. Another class of methods involves a template curve, to which each observation is registered, minimising a discrepancy criterion; the template is then iteratively updated (Wang and Gasser [129]; Ramsay and Li [108]). James [72] defines a "feature function" for each curve and uses the moments of the feature function to guarantee identifiability. Elastic registration employs the Fisher–Rao metric that is invariant to warpings and calculates averages in the resulting quotient space (Tucker et al. [123]). Other techniques include semiparametric modelling (Rønn [115]; Gervini and Gasser [64]) and principal components registration (Kneip and Ramsay [82]). More details can be found in the review article by Marron et al. [90]. Wrobel et al. [131] have recently developed a registration method for functional data with a discrete flavour. It is also noteworthy that a version of the Wasserstein metric can also be used in the functional case (Chakraborty and Panaretos [34]).

The literature on the point processes case is more scarce; see the review by Wu and Srivastava [133].

A parametric version of Theorem 4.2.4 was first established by Bigot and Klein [22, Theorem 5.1] in $\mathbb{R}^d$, extended to a compact nonparametric formulation in Zemel and Panaretos [134]. There is an infinite-dimensional linear version in Masarotto et al. [91]. The current level of generality appears to be new.

Theorem 4.4.1 is a stronger version of Panaretos and Zemel [100, Theorem 1] where it was assumed that τ_n must diverge to infinity faster than $\log n$. An analogous construction under the Bayesian paradigm can be found in Galasso et al. [58]. Optimality of the rates of convergence in Theorem 4.6.3 is discussed in detail by Bigot et al. [23], where finiteness of the functional J_2 (see Sect. 4.7) is assumed and consequently $O_{\mathbb{P}}(\tau_n^{-1/4})$ is improved to $O_{\mathbb{P}}(\tau_n^{-1/2})$.

As far as we know, Theorem 4.6.5 (taken from [100]) is the first central limit theorem for Fréchet means in Wasserstein space. When the measures Λ_i are observed exactly (no amplitude variation: $\tau_n = \infty$ and $\sigma = 0$) Kroshnin et al. [84] have recently proven a central limit theorem for random Gaussian measures in arbitrary dimension, extending a previous result of Agueh and Carlier [3]. It seems likely that in a fully nonparametric setting, the rates of convergence (compare Theorem 4.6.3) might be slower than $\sqrt{n}$; see Ahidar-Coutrix et al. [4].

The magnitude of the amplitude variation in Theorem 4.6.3 pertains to the rates of convergence of $\mathbb{E}W_p(\mu_n,\mu)$ to zero (Sect. 4.7). This is a topic of intense research, dating back to the seminal paper by Dudley [46], where a version of Theorem 4.7.8 with $p=1$ is shown for the bounded Lipschitz metric. The lower bounds proven in this section were adapted from [46], Fournier and Guillin [54], and Weed and Bach [130].

The version of Theorem 4.7.8 given here can be found in [130] and extends Boissard and Le Gouic [27]. Both papers [27, 130] work in a general setting of complete separable metric spaces. An additional $\log n$ term appears in the limiting case $d=2p$, as already noted (for $p=1$) by [46], and the classical work of Ajtai et al. [5] for μ uniform on $[0,1]^2$. More general results are available in [54]. A longer (but far from being complete) bibliography is given in the recent review by Panaretos and Zemel [101, Subsection 3.3.1], including works by Barthe, Dobrić, Talagrand, and coauthors on almost sure results and deviation bounds for the empirical Wasserstein distance.

The J_1 condition is due to del Barrio et al. [43], who showed it to be necessary and sufficient for the empirical process $\sqrt{n}(F_n-F)$ to converge in distribution to $\mathbb{B}\circ F$, with $\mathbb{B}$ Brownian bridge. The extension to $1\le p\le\infty$ (and a lot more) can be found in Bobkov and Ledoux [25], employing order statistics and beta distributions to reduce to the uniform case. Alternatively, one may consult Mason [92], who uses weighted approximations to Brownian bridges.

An important aspect that was not covered here is that of statistical inference of the Wasserstein distance on the basis of the empirical measure. This is a challenging question and results by del Barrio, Munk, and coauthors are available for one-dimensional, elliptical, or discrete measures, as explained in [101, Section 3].

Chapter 5
Construction of Fréchet Means and Multicouplings

When given measures $\mu^1,\dots,\mu^N$ are supported on the real line, computing their Fréchet mean $\bar{\mu}$ is straightforward (Sect. 3.1.4). This is in contrast to the multivariate case, where, apart from the important yet special case of compatible measures, closed-form formulae are not available. This chapter presents an iterative procedure that provably approximates at least a Karcher mean with mild restrictions on the measures $\mu^1,\dots,\mu^N$. The algorithm is based on the differentiability properties of the Fréchet functional developed in Sect. 3.1.6 and can be interpreted as classical steepest descent in the Wasserstein space $\mathscr{W}_2(\mathbb{R}^d)$. It reduces the problem of finding the Fréchet mean to a succession of pairwise transport problems, involving only the Monge–Kantorovich problem between two measures. In the Gaussian case (or any location-scatter family), the latter can be done explicitly, rendering the algorithm particularly appealing (see Sect. 5.4.1).

This chapter can be seen as a complementary to Chap. 4. On the one hand, one can use the proposed algorithm to construct the regularised Fréchet–Wasserstein estimator $\widehat{\lambda}_n$ that approximates a population version (see Sect. 4.3). On the other hand, it could be that the object of interest is the sample $\mu^1,\dots,\mu^N$ itself, but that the latter is observed with some amount of noise. If one only has access to proxies $\widehat{\mu^1},\dots,\widehat{\mu^N}$, then it is natural to use their Fréchet mean $\widehat{\bar{\mu}}$ as an estimator of $\bar{\mu}$. The proposed algorithm can then be used, in principle, in order to construct $\bar{\mu}$, and the consistency framework of Sect. 4.4 then allows to conclude that if each $\widehat{\mu^i}$ is consistent, then so is $\widehat{\bar{\mu}}$.

After presenting the algorithm in Sect. 5.1, we make some connections to Procrustes analysis in Sect. 5.2. A convergence analysis of the algorithm is carried out in Sect. 5.3, after which examples are given in Sect. 5.4. An extension to infinitely many measures is sketched in Sect. 5.5.

Electronic Supplementary Material The online version of this chapter (https://doi.org/10.1007/978-3-030-38438-8_5) contains supplementary material.

V. M. Panaretos, Y. Zemel, *An Invitation to Statistics in Wasserstein Space*, SpringerBriefs in Probability and Mathematical Statistics,
https://doi.org/10.1007/978-3-030-38438-8_5

5.1 A Steepest Descent Algorithm for the Computation of Fréchet Means

Throughout this section, we assume that N is a fixed integer and consider a fixed collection

$$\mu^1,\dots,\mu^N \in \mathscr{W}_2(\mathbb{R}^d) \qquad \text{with } \mu^1 \text{ absolutely continuous with bounded density,} \tag{5.1}$$

whose unique (Proposition 3.1.8) Fréchet mean $\bar{\mu}$ is sought. It has been established that if γ is absolutely continuous then the associated Fréchet functional

$$F(\gamma) = \frac{1}{2N}\sum_{i=1}^{n} W_2^2(\mu^i,\gamma), \qquad \gamma \in \mathscr{W}_2(\mathbb{R}^d),$$

has Fréchet derivative (Theorem 3.1.14)

$$F'(\gamma) = -\frac{1}{N}\sum_{i=1}^{N}\log_\gamma(\mu^i) = -\frac{1}{N}\sum_{i=1}^{N}\left(\mathbf{t}_\gamma^{\mu_i} - \mathbf{i}\right) \in \mathrm{Tan}_\gamma \tag{5.2}$$

at γ. Let $\gamma_j \in \mathscr{W}_2(\mathbb{R}^d)$ be an absolutely continuous measure, representing our current estimate of the Fréchet mean at step j. Then it makes sense to introduce a step size $\tau_j > 0$, and to follow the steepest descent of F given by the negative of the gradient:

$$\gamma_{j+1} = \exp_{\gamma_j}\left(-\tau_j F'(\gamma_j)\right) = \left[\mathbf{i} + \tau_j\frac{1}{N}\sum_{i=1}^{N}\log_\gamma(\mu^i)\right]\#\gamma_j = \left[\mathbf{i} + \tau_j\frac{1}{N}\sum_{i=1}^{N}(\mathbf{t}_{\gamma_j}^{\mu^i} - \mathbf{i})\right]\#\gamma_j.$$

In order to employ further descent at γ_{j+1}, it needs to be verified that F is differentiable at γ_{j+1}, which amounts to showing that the latter stays absolutely continuous. This will happen for all but countably many values of the step size τ_j, but necessarily if the latter is contained in $[0,1]$:

Lemma 5.1.1 (Regularity of the Iterates) *If γ_0 is absolutely continuous and $\tau = \tau_0 \in [0,1]$, then $\gamma_1 = \exp_{\gamma_0}(-\tau_0 F'(\gamma_0))$ is also absolutely continuous.*

The idea is that push-forwards of γ_0 under monotone maps are absolutely continuous if and only if the monotonicity is strict, a property preserved by averaging. See page 118 in the supplement for the details.

Lemma 5.1.1 suggests that the step size should be restricted to $[0,1]$. The next result suggests that the objective function essentially tells us that the optimal step size, achieving the maximal reduction of the objective function (thus corresponding to an approximate line search), is exactly equal to 1. It does not rely on finite-dimensional arguments and holds when replacing $\mathbb{R}^d$ by a separable Hilbert space.

Lemma 5.1.2 (Optimal Stepsize) *If $\gamma_0 \in \mathscr{W}_2(\mathbb{R}^d)$ is absolutely continuous, then*

$$F(\gamma_1) - F(\gamma_0) \leq -\|F'(\gamma_0)\|^2 \left[\tau - \frac{\tau^2}{2}\right]$$

and the bound on the right-hand side of the last display is minimised when $\tau = 1$.

Proof. Let $S_i = \mathbf{t}_{\gamma_0}^{\mu^i}$ be the optimal map from γ_0 to μ^i, and set $W_i = S_i - \mathbf{i}$. Then

$$2NF(\gamma_0) = \sum_{i=1}^{N} W_2^2(\gamma_0, \mu^i) = \sum_{i=1}^{N} \int_{\mathbb{R}^d} \|S_i - \mathbf{i}\|^2 \, d\gamma_0 = \sum_{i=1}^{N} \|W_i\|^2_{\mathscr{L}^2(\gamma_0)}, \tag{5.3}$$

Both γ_1 and μ^i can be written as push-forwards of γ_0 and (2.3) gives the bound

$$W_2^2(\gamma_1, \mu^i) \leq \int_{\mathbb{R}^d} \left\| \left[(1-\tau)\mathbf{i} + \frac{\tau}{N}\sum_{j=1}^{N} S_j\right] - S_i \right\|^2_{\mathbb{R}^d} d\gamma_0 = \left\| -W_i + \frac{\tau}{N}\sum_{j=1}^{N} W_j \right\|^2_{\mathscr{L}^2(\gamma_0)}.$$

For brevity, we omit the subscript $\mathscr{L}^2(\gamma_0)$ from the norms and inner products. Developing the squares, summing over $i = 1, \ldots, N$ and using (5.3) gives

$$\begin{aligned} 2NF(\gamma_1) &\leq \sum_{i=1}^{N} \|W_i\|^2 - 2\frac{\tau}{N} \sum_{i,j=1}^{N} \langle W_i, W_j \rangle + N\tau^2 \left\| \sum_{j=1}^{N} \frac{1}{N} W_j \right\|^2 \\ &= 2NF(\gamma_0) - 2N\tau \left\| \sum_{i=1}^{N} \frac{1}{N} W_i \right\|^2 + N\tau^2 \left\| \sum_{i=1}^{N} \frac{1}{N} W_i \right\|^2, \end{aligned}$$

and recalling that $W_i = S_i - \mathbf{i}$ yields

$$F(\gamma_1) - F(\gamma_0) \leq \frac{\tau^2 - 2\tau}{2} \left\| \frac{1}{N} \sum_{i=1}^{N} W_i \right\|^2 = -\|F'(\gamma_0)\|^2 \left[\tau - \frac{\tau^2}{2}\right].$$

To conclude, observe that $\tau - \tau^2/2$ is maximised at $\tau = 1$.

In light of Lemmata 5.1.1 and 5.1.2, we will always take $\tau_j = 1$. The resulting iteration is summarised as Algorithm 1. A first step in the convergence analysis is that the sequence $(F(\gamma_j))$ is nonincreasing and that for any integer k,

$$\frac{1}{2} \sum_{j=0}^{k} \|F'(\gamma_j)\|^2 \leq \sum_{j=0}^{k} F(\gamma_j) - F(\gamma_{j+1}) = F(\gamma_0) - F(\gamma_{k+1}) \leq F(\gamma_0).$$

As $k \to \infty$, the infinite sum on the left-hand side converges, so $\|F'(\gamma_j)\|^2$ must vanish as $j \to \infty$.

Remark 5.1.3 *The proof of Proposition 3.1.2 suggests a generalisation of Algorithm 1 to arbitrary measures in $\mathcal{W}_2(\mathbb{R}^d)$ even if none are absolutely continuous. One can verify that Lemmata 5.1.2 and 5.3.5 (below) also hold in this setup, so it may be that convergence results also apply in this setup. The iteration no longer has the interpretation as steepest descent, however.*

Algorithm 1 Steepest descent via Procrustes analysis

(A) Set a tolerance threshold $\varepsilon > 0$.
(B) For $j = 0$, let γ_j be an arbitrary absolutely continuous measure.
(C) For $i = 1, \dots, N$ solve the (pairwise) Monge problem and find the optimal transport map $\mathbf{t}_{\gamma_j}^{\mu^i}$ from γ_j to μ^i.
(D) Define the map $T_j = N^{-1} \sum_{i=1}^{N} \mathbf{t}_{\gamma_j}^{\mu^i}$.
(E) Set $\gamma_{j+1} = T_j \# \gamma_j$, i.e. push-forward γ_j via T_j to obtain γ_{j+1}.
(F) If $\|F'(\gamma_{j+1})\| < \varepsilon$, stop, and output γ_{j+1} as the approximation of $\bar{\mu}$ and $\mathbf{t}_{\gamma_{j+1}}^{\mu^i}$ as the approximation of $\mathbf{t}_{\bar{\mu}}^{\mu^i}$, $i = 1, \dots, N$. Otherwise, return to step (C).

5.2 Analogy with Procrustes Analysis

Algorithm 1 is similar in spirit to another procedure, *generalised Procrustes analysis*, that is used in shape theory. Given a subset $B \subseteq \mathbb{R}^d$, most commonly a finite collection of labelled points called *landmarks*, an interesting question is how to mathematically define the *shape* of B. One way to reach such a definition is to disregard those properties of B that are deemed irrelevant for what one considers this shape should be; typically, these would include its location, its orientation, and/or its scale. Accordingly, the shape of B can be defined as the equivalence class consisting of all sets obtained as gB, where g belongs to a collection $\mathcal{G}$ of transformations of $\mathbb{R}^d$ containing all combinations of rotations, translations, dilations, and/or reflections (Dryden and Mardia [45, Chapter 4]).

If B_1 and B_2 are two collections of k landmarks, one may define the distance between their shapes as the infimum of $\|B_1 - gB_2\|^2$ over the group $\mathcal{G}$. In other words, one seeks to *register* B_2 as close as possible to B_1 by using elements of the group $\mathcal{G}$, with distance being measured as the sum of squared Euclidean distances between the transformed points of B_2 and those of B_1. In a sense, one can think about the shape problem and the Monge problem as dual to each other. In the former, one is given constraints on how to optimally carry out the registration of the points with the cost being judged by how successful the registration procedure is. In the latter, one imposes that the registration be done *exactly*, and evaluates the cost by how much the space must be deformed in order to achieve this.

The optimal g and the resulting distance can be found in closed-form by means of *ordinary Procrustes analysis* [45, Section 5.2]. Suppose now that we are given

$N > 2$ collections of points, $B_1,\dots,B_N$, with the goal of minimising the sum of squares $\|g_iB_i - g_jB_j\|^2$ over $g_i \in \mathscr{G}$.[1] As in the case of Fréchet means in $\mathscr{W}_2(\mathbb{R}^d)$ (Sect. 3.1.2), there is a formulation in terms of sum of squares from the average $N^{-1}\sum g_jB_j$. Unfortunately, there is no explicit solution for this problem when $d \geq 3$. Like Algorithm 1, generalised Procrustes analysis (Gower [66]; Dryden and Mardia [45, p. 90]) tackles this "multimatching" setting by iteratively solving the pairwise problem as follows. Choose one of the configurations as an initial estimate/template, then register every other configuration to the template, employing ordinary Procrustes analysis. The new template is then given by the linear average of the registered configurations, and the process is iterated subsequently.

Paralleling this framework, Algorithm 1 iterates the two steps of registration and linear averaging given the current template γ_j, but in a different manner:

(1) **Registration**: by finding the optimal transportation maps $\mathbf{t}_{\gamma_j}^{\mu^i}$, we identify each μ^i with the element $\mathbf{t}_{\gamma_j}^{\mu^i} - \mathbf{i} = \log_{\gamma_j}(\mu^i)$. In this sense, the collection $(\mu^1,\dots,\mu^N)$ is viewed in the common coordinate system given by the tangent space at the template γ_j and is registered to it.
(2) **Averaging**: the registered measures are averaged linearly, using the common coordinate system of the registration step (1), as elements in the linear space Tan_{γ_j}. The linear average is then retracted back onto the Wasserstein space via the exponential map to yield the estimate at the $(j+1)$-th step, γ_{j+1}.

Notice that in the Procrustes sense, the maps that register each μ^i to the template γ_j are $\mathbf{t}_{\mu^i}^{\gamma_j}$, the inverses of $\mathbf{t}_{\gamma_j}^{\mu^i}$. We will not use the term "registration maps" in the sequel, to avoid possible confusion.

5.3 Convergence of Algorithm 1

In order to tackle the issue of convergence, we will use an approach that is specific to the nature of optimal transport. This is because the Hessian-type arguments that are used to prove similar convergence results for steepest descent on Riemannian manifolds (Afsari et al. [1]) or Procrustes algorithms (Le [86]; Groisser [67]) do not apply here, since the Fréchet functional may very well fail to be twice differentiable.

In fact, even in Euclidean spaces, convergence of steepest descent usually requires a Lipschitz bound on the derivative of F (Bertsekas [19, Subsection 1.2.2]). Unfortunately, F is not known to be differentiable at discrete measures, and these constitute a dense set in $\mathscr{W}_2$; consequently, this Lipschitz condition is very unlikely to hold. Still, this specific geometry of the Wasserstein space affords some advantages; for instance, we will place no restriction on the starting point for the iteration, except that it be absolutely continuous; and no assumption on how "spread out" the collection $\mu^1,\dots,\mu^N$ is will be necessary as in, for example, [1, 67, 86].

---

[1] One needs to add an additional constraint to prevent registering all the collections to the origin.

Theorem 5.3.1 (Limit Points are Karcher Means) *Let $\mu^1,\dots,\mu^N \in \mathcal{W}_2(\mathbb{R}^d)$ be probability measures and suppose that one of them is absolutely continuous with a bounded density. Then, the sequence generated by Algorithm 1 stays in a compact set of the Wasserstein space $\mathcal{W}_2(\mathbb{R}^d)$, and any limit point of the sequence is a Karcher mean of $(\mu^1,\dots,\mu^N)$.*

Since the Fréchet mean $\bar{\mu}$ is a Karcher mean (Proposition 3.1.8), we obtain immediately:

Corollary 5.3.2 (Wasserstein Convergence of Steepest Descent) *Under the conditions of Theorem 5.3.1, if F has a unique stationary point, then the sequence $\{\gamma_j\}$ generated by Algorithm 1 converges to the Fréchet mean of $\{\mu^1,\dots,\mu^N\}$ in the Wasserstein metric,*

$$W_2(\gamma_j,\bar{\mu})\longrightarrow 0, \qquad j\to\infty.$$

Alternatively, combining Theorem 5.3.1 with the optimality criterion Theorem 3.1.15 shows that the algorithm converges to $\bar{\mu}$ when the appropriate assumptions on $\{\mu^i\}$ and the Karcher mean $\mu = \lim \gamma_j$ are satisfied. This allows to conclude that Algorithm 1 converges to the unique Fréchet mean when μ^i are Gaussian measures (see Theorem 5.4.1).

The proof of Theorem 5.3.1 is rather elaborate, since we need to use specific methods that are tailored to the Wasserstein space. Before giving the proof, we state two important consequences. The first is the uniform convergence of the optimal maps $\mathbf{t}_{\gamma_j}^{\mu^i}$ to $\mathbf{t}_{\bar{\mu}}^{\mu^i}$ on compacta. This convergence does not immediately follow from the Wasserstein convergence of γ_j to $\bar{\mu}$, and is also established for the inverses. Both the formulation and the proof of this result are similar to those of Theorem 4.4.3.

Theorem 5.3.3 (Uniform Convergence of Optimal Maps) *Under the conditions of Corollary 5.3.2, there exist sets $A, B^1,\dots,B^N \subseteq \mathbb{R}^d$ such that $\bar{\mu}(A) = 1 = \mu^1(B^1) = \cdots = \mu^N(B^N)$ and*

$$\sup_{\Omega_1}\left\|\mathbf{t}_{\gamma_j}^{\mu^i} - \mathbf{t}_{\bar{\mu}}^{\mu^i}\right\| \stackrel{j\to\infty}{\longrightarrow} 0, \qquad \sup_{\Omega_2^i}\left\|\mathbf{t}_{\mu^i}^{\gamma_j} - \mathbf{t}_{\mu^i}^{\bar{\mu}}\right\| \stackrel{j\to\infty}{\longrightarrow} 0, \qquad i=1,\dots,N,$$

for any pair of compacta $\Omega_1 \subseteq A$, $\Omega_2^i \subseteq B^i$. If in addition all the measures $\mu^1,\dots,\mu^N$ have the same support, then one can choose all the sets B^i to be the same.

The other consequence is convergence of the optimal multicouplings.

Corollary 5.3.4 (Convergence of Multicouplings) *Under the conditions of Corollary 5.3.2, the sequence of multicouplings*

$$\left(\mathbf{t}_{\gamma_j}^{\mu^1},\dots\mathbf{t}_{\gamma_j}^{\mu^n}\right)\#\gamma_j$$

of $\{\mu^1,\dots,\mu^N\}$ converges (in Wasserstein distance on $(\mathbb{R}^d)^N$) to the optimal multicoupling $(\mathbf{t}_{\overline{\mu}}^{\mu^1},\dots\mathbf{t}_{\overline{\mu}}^{\mu^n})\#\overline{\mu}$.

The proofs of Theorem 5.3.3 and Corollary 5.3.4 are given at the end of the present section.

The proof of Theorem 5.3.1 is achieved by establishing the following facts:

1. The sequence (γ_j) stays in a compact subset of $\mathscr{W}_2(\mathbb{R}^d)$ (Lemma 5.3.5).
2. Any limit of (γ_j) is absolutely continuous (Proposition 5.3.6 and the paragraph preceding it).
3. Algorithm 1 acts continuously on its argument (Corollary 5.3.8).

Since it has already been established that $\|F'(\gamma_j)\| \to 0$, these three facts indeed suffice.

Lemma 5.3.5 *The sequence generated by Algorithm 1 stays in a compact subset of the Wasserstein space $\mathscr{W}_2(\mathbb{R}^d)$.*

Proof. For all $j \geq 1$, γ_j takes the form $M_n\#\pi$, where $M_N(x_1,\dots,x_N) = \overline{x}$ and π is a multicoupling of $\mu^1,\dots,\mu^N$. The compactness of this set has been established in Step 2 of the proof of Theorem 3.1.5; see page 63 in the supplement, where this is done in a more complicated setup.

A closer look at the proof reveals that a more general result holds true. Let $\mathscr{A}$ denote the steepest descent iteration, that is, $\mathscr{A}(\gamma_j) = \gamma_{j+1}$. Then the image of $\mathscr{A}$, $\{\mathscr{A}\mu : \mu \in \mathscr{W}_2(\mathbb{R}^d) \text{ absolutely continuous}\}$ has a compact closure in $\mathscr{W}_2(\mathbb{R}^d)$. This is also true if $\mathbb{R}^d$ is replaced by a separable Hilbert space.

In order to show that a weakly convergent sequence (γ_j) of absolutely continuous measures has an absolutely continuous limit γ, it suffices to show that the densities of γ_j are uniformly bounded. Indeed, if C is such a bound, then for any open $O \subseteq \mathbb{R}^d$, $\liminf \gamma_k(O) \leq C\text{Leb}(O)$, so $\gamma(O) \leq C\text{Leb}(O)$ by the portmanteau Lemma 1.7.1. It follows that γ is absolutely continuous with density bounded by C. We now show that such C can be found that applies to all measures in the image of $\mathscr{A}$, hence to all sequences resulting from iterations of Algorithm 1.

Proposition 5.3.6 (Uniform Density Bound) *For each $i = 1,\dots,N$ denote by g^i the density of μ^i (if it exists) and $\|g^i\|_\infty$ its supremum, taken to be infinite if g^i does not exist (or if g^i is unbounded). Let γ_0 be any absolutely continuous probability measure. Then the density of $\gamma_1 = \mathscr{A}(\gamma_0)$ is bounded by the $1/d$-th harmonic mean of $\|g^i\|_\infty$,*

$$C_\mu = \left[\frac{1}{N}\sum_{i=1}^{N}\frac{1}{\|g^i\|_\infty^{1/d}}\right]^{-d}.$$

The constant C_μ depends only on the measures $(\mu^1,\dots,\mu^N)$, and is finite as long as one μ^i has a bounded density, since $C_\mu \leq N^d\|g^i\|_\infty$ for any i.

Proof. Let h_i be the density of γ_i. By the change of variables formula, for γ_0-almost any x

$$h_1(\mathbf{t}_{\gamma_0}^{\gamma_1}(x)) = \frac{h_0(x)}{\det \nabla \mathbf{t}_{\gamma_0}^{\gamma_1}(x)}; \qquad g^i(\mathbf{t}_{\gamma_0}^{\mu^i}(x)) = \frac{h_0(x)}{\det \nabla \mathbf{t}_{\gamma_0}^{\mu^i}(x)}, \qquad \text{when } g^i \text{ exists.}$$

(Convex functions are twice differentiable almost surely (Villani [125, Theorem 14.25]), hence these gradients are well-defined γ_0-almost surely.) We seek a lower bound on the determinant of $\nabla \mathbf{t}_{\gamma_0}^{\gamma_1}(x)$, which by definition equals

$$N^{-d} \det \sum_{i=1}^{N} \nabla \mathbf{t}_{\gamma_0}^{\mu^i}(x).$$

Such a bound is provided by the Brunn–Minkowski inequality (Stein and Shakarchi [121, Section 1.5]) for symmetric positive semidefinite matrices

$$[\det(A+B)]^{1/d} \geq [\det A]^{1/d} + [\det B]^{1/d},$$

which, applied inductively, yields

$$\left[\det \nabla \mathbf{t}_{\gamma_0}^{\gamma_1}(x)\right]^{1/d} \geq \frac{1}{N} \sum_{i=1}^{N} \left[\det \nabla \mathbf{t}_{\gamma_0}^{\mu^i}(x)\right]^{1/d}.$$

From this, we obtain an upper bound for h_1:

$$\frac{1}{h_1^{1/d}(\mathbf{t}_{\gamma_0}^{\gamma_1}(x))} = \frac{\det^{1/d} \sum_{i=1}^{N} \nabla \mathbf{t}_{\gamma_0}^{\mu^i}(x)}{N h_0^{1/d}(x)} \geq \frac{1}{N} \sum_{i=1}^{N} \frac{1}{[g^i(\mathbf{t}_{\gamma_0}^{\mu^i}(x))]^{1/d}} \geq \frac{1}{N} \sum_{i=1}^{N} \frac{1}{\|g^i\|_\infty^{1/d}} = C_\mu^{-1/d}.$$

Let Σ be the set of points where this inequality holds, then $\gamma_0(\Sigma) = 1$. Hence

$$\gamma_1(\mathbf{t}_{\gamma_0}^{\gamma_1}(\Sigma)) = \gamma_0[(\mathbf{t}_{\gamma_0}^{\gamma_1})^{-1}(\mathbf{t}_{\gamma_0}^{\gamma_1}(\Sigma))] \geq \gamma_0(\Sigma) = 1.$$

Thus, γ_1-almost surely and for all i,

$$h_1(y) \leq C_\mu.$$

The third statement (continuity of $\mathscr{A}$) is much more subtle to establish, and its rather lengthy proof is given next. In view of Proposition 5.3.6, the uniform bound on the densities is not a hindrance for the proof of convergence of Algorithm 1.

Proposition 5.3.7 *Let (γ_n) be a sequence of absolutely continuous measures with uniformly bounded densities, suppose that $W_2(\gamma_n, \gamma) \to 0$, and let*

$$\eta_j = \left(\mathbf{t}_{\gamma_j}^{\mu^1}, \ldots \mathbf{t}_{\gamma_j}^{\mu^n}, \mathbf{i}\right) \# \gamma_j, \qquad \eta = \left(\mathbf{t}_{\gamma}^{\mu^1}, \ldots \mathbf{t}_{\gamma}^{\mu^n}, \mathbf{i}\right) \# \gamma.$$

Then $\eta_j \to \eta$ in $\mathscr{W}_2([\mathbb{R}^d]^{N+1})$.

Proof. As has been established in the discussion before Proposition 5.3.6, the limit γ must be absolutely continuous, so η is well-defined.

In view of Theorem 2.2.1, it suffices to show that if $h : (\mathbb{R}^d)^{N+1} \to \mathbb{R}$ is any continuous nonnegative function such that

$$|h(t_1,\dots,t_N,y)| \le \frac{2}{N}\sum_{i=1}^{N}\|t_i\|^2 + 2\|y\|^2,$$

then

$$\int_{\mathbb{R}^d} g_n \, d\gamma_n = \int_{(\mathbb{R}^d)^{N+1}} h \, d\eta_n \to \int_{(\mathbb{R}^d)^{N+1}} h \, d\eta = \int_{\mathbb{R}^d} g \, d\gamma, \; g_n(x) = h(\mathbf{t}_{\gamma_j}^{\mu^1}(x), \dots \mathbf{t}_{\gamma_j}^{\mu^n}(x), x),$$

and g defined analogously. The proof, given in full detail on page 124 of the supplement, is sketched here.

Step 1: Truncation. Since γ_n converge in the Wasserstein space, they satisfy the uniform integrability (2.4) and absolute continuity (2.7) by Theorem 2.2.1. Consequently, $g_{n,R} = \min(g_n, 4R)$ is uniformly close to g_n:

$$\sup_n \int [g_n(x) - g_{n,R}(x)] \, d\gamma_n(x) \to 0, \qquad R \to \infty.$$

We may thus replace g_n by a bounded version $g_{n,R}$.

Step 2: Convergence of g_n to g. By Proposition 1.7.11, the optimal maps $\mathbf{t}_{\gamma_n}^{\mu^i}$ converge to $\mathbf{t}_{\gamma}^{\mu^i}$ and (since h is continuous), $g_n \to g$ uniformly on "nice" sets $\Omega \subseteq E = \operatorname{supp}\gamma$. Write

$$\int g_{n,R} \, d\gamma_n - \int g_R \, d\gamma = \int g_R \, d(\gamma_n - \gamma) + \int_{\Omega} (g_{n,R} - g_R) \, d\gamma_n + \int_{\mathbb{R}^d \setminus \Omega} (g_{n,R} - g_R) \, d\gamma_n.$$

Step 3: Bounding the first two integrals. The first integral vanishes as $n \to \infty$, by the portmanteau Lemma 1.7.1, and the second by uniform convergence.

Step 4: Bounding the third integral. The integrand is bounded by $8R$, so it suffices to bound the measures of $\mathbb{R}^d \setminus \Omega$. This is a bit technical, and uses the uniform density bound on (γ_n) and the portmanteau lemma.

Corollary 5.3.8 (Continuity of $\mathscr{A}$) *If $W_2(\gamma_n, \gamma) \to 0$ and γ_n have uniformly bounded densities, then $\mathscr{A}(\gamma_n) \to \mathscr{A}(\gamma)$.*

Proof. Choose h in the proof of Proposition 5.3.7 to depend only on y.

Proof (Proof of Corollary 5.3.4). Choose h in the proof of Proposition 5.3.7 to depend only on $t_1, \dots, t_n$.

Proof (Proof of Theorem 5.3.3). Let $E = \operatorname{supp}\bar{\mu}$ and set $A^i = E^{\text{den}} \cap \{x : \mathbf{t}_{\bar{\mu}}^{\mu^i}(x) \text{ is univalued}\}$. As $\bar{\mu}$ is absolutely continuous, $\bar{\mu}(A^i) = 1$, and the same is true for $A = \cap_{i=1}^N A^i$. The first assertion then follows from Proposition 1.7.11.

The second statement is proven similarly. Let $E^i = \operatorname{supp}\mu^i$ and notice that by absolute continuity the $B^i = (E^i)^{\text{den}} \cap \{x : \mathbf{t}_{\mu^i}^{\bar{\mu}}(x) \text{ is univalued}\}$ has measure 1 with respect to μ^i. Apply Proposition 1.7.11. If in addition $E^1 = \dots = E^N$, then $\mu^i(B) = 1$ for $B = \cap B^i$.

5.4 Illustrative Examples

As an illustration, we implement Algorithm 1 in several scenarios for which pairwise optimal maps can be calculated explicitly at every iteration, allowing for fast computation without error propagation. In each case, we give some theory first, describing how the optimal maps are calculated, and then implement Algorithm 1 on simulated examples.

5.4.1 Gaussian Measures

No example illustrates the use of Algorithm 1 better than the Gaussian case. This is so because optimal maps between centred nondegenerate Gaussian measures with covariances A and B have the explicit form (see Sect. 1.6.3)

$$\mathbf{t}_A^B(x) = A^{-1/2}[A^{1/2}BA^{1/2}]^{1/2}A^{-1/2}x, \qquad x \in \mathbb{R}^d,$$

with the obvious slight abuse of notation. In contrast, the Fréchet mean of a collection of Gaussian measures (one of which nonsingular) does not admit a closed-form formula and is only known to be a Gaussian measure whose covariance matrix Γ is the unique invertible root of the matrix equation

$$\Gamma = \frac{1}{N}\sum_{i=1}^{N}\left[\Gamma^{1/2}S_i\Gamma^{1/2}\right]^{1/2}, \tag{5.4}$$

where S_i is the covariance matrix of μ^i.

Given the formula for $\mathbf{t}_A^B$, application of Algorithm 1 to Gaussian measures is straightforward. The next result shows that, in the Gaussian case, the iterates must converge to the unique Fréchet mean, and that (5.4) can be derived from the characterisation of Karcher means.

Theorem 5.4.1 (Convergence in Gaussian Case) *Let $\mu^1,\dots,\mu^N$ be Gaussian measures with zero means and covariance matrices S_i with S_1 nonsingular, and let the initial point γ_0 be $\mathcal{N}(0,\Gamma_0)$ with Γ_0 nonsingular. Then the sequence of iterates generated by Algorithm 1 converges to the unique Fréchet mean of $(\mu^1,\dots,\mu^N)$.*

Proof. Since the optimal maps are linear, so is their mean and therefore γ_k is a Gaussian measure for all k, say $\mathcal{N}(0,\Gamma_k)$ with Γ_k nonsingular. Any limit point of γ is a Karcher mean by Theorem 5.3.1. If we knew that γ itself were Gaussian, then it actually must be the Fréchet mean because $N^{-1}\sum\mathbf{t}_\gamma^{\mu^i}$ equals the identity everywhere on $\mathbb{R}^d$ (see the discussion before Theorem 3.1.15).

Let us show that every limit point γ is indeed Gaussian. It suffices to prove that (Γ_k) is a bounded sequence, because if $\Gamma_k \to \Gamma$, then $\mathcal{N}(0,\Gamma_k) \to \mathcal{N}(0,\Gamma)$ weakly, as can be seen from either Lehmann–Scheffé's theorem (the densities converge) or Lévy's continuity theorem (the characteristic functions converge).

To see that (Γ_k) is bounded, observe first that for any centred (Gaussian or not) measure μ with covariance matrix S,

$$W_2^2(\mu,\delta_0) = \mathrm{tr}[S],$$

where δ_0 is a Dirac mass at the origin. (This follows from the spectral decomposition of S.) Therefore

$$0 \le \mathrm{tr}[\Gamma_k] = W_2^2(\gamma_k,\delta_0)$$

is bounded uniformly, because $\{\gamma_k\}$ stays in a Wasserstein-compact set by Lemma 5.3.5. If we define $C = \sup_k \mathrm{tr}[\Gamma_k] < \infty$, then all the diagonal elements of Γ_k are bounded uniformly. When A is symmetric and positive semidefinite, $2|A_{ij}| \le A_{ii} + A_{ij}$. Consequently, all the entries of Γ_k are bounded uniformly by C, which means that (Γ_k) is a bounded sequence.

From the formula for the optimal maps, we see that if Γ is the covariance of the Fréchet mean, then

$$I = \frac{1}{N}\sum_{i=1}^{N} \Gamma^{-1/2}\left[\Gamma^{1/2} S_i \Gamma^{1/2}\right]^{1/2}\Gamma^{-1/2}$$

and we recover the fixed point equation (5.4).

If the means are nonzero, then the optimal maps are affine and the same result applies; the Fréchet mean is still a Gaussian measure with covariance matrix Γ and mean that equals the average of the means of μ^i, $i = 1,\dots,N$.

Figure 5.1 shows density plots of $N = 4$ centred Gaussian measures on $\mathbb{R}^2$ with covariances $S_i \sim \mathrm{Wishart}(I_2, 2)$, and Fig. 5.2 shows the density of the resulting Fréchet mean. In this particular example, the algorithm needed 11 iterations starting from the identity matrix. The corresponding optimal maps are displayed in Fig. 5.3. It is apparent from the figure that these maps are linear, and after a more careful reflection one can be convinced that their average is the identity. The four plots in the figure are remarkably different, in accordance with the measures themselves having widely varying condition numbers and orientations; μ^3 and more so μ^4 are very concentrated, so the optimal maps "sweep" the mass towards zero. In contrast, the optimal maps to μ^1 and μ^2 spread the mass out away from the origin.

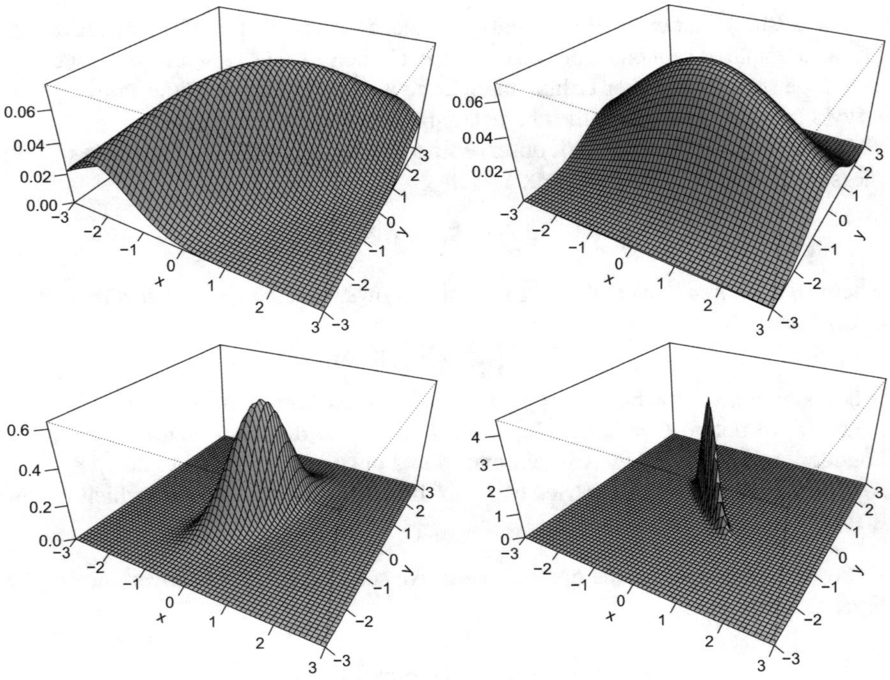

Fig. 5.1: Density plot of four Gaussian measures in $\mathbb{R}^2$.

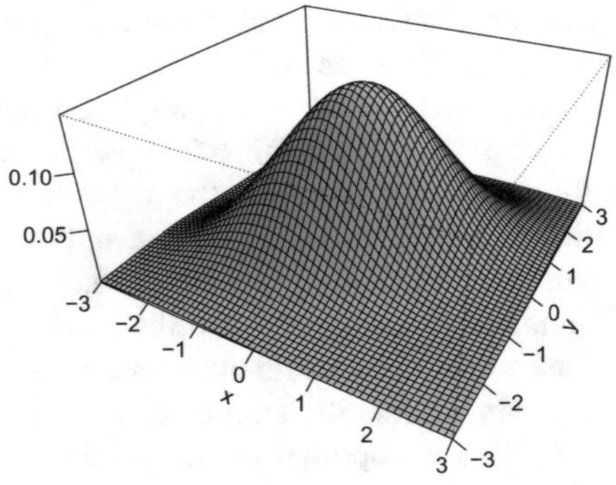

Fig. 5.2: Density plot of the Fréchet mean of the measures in Fig. 5.1

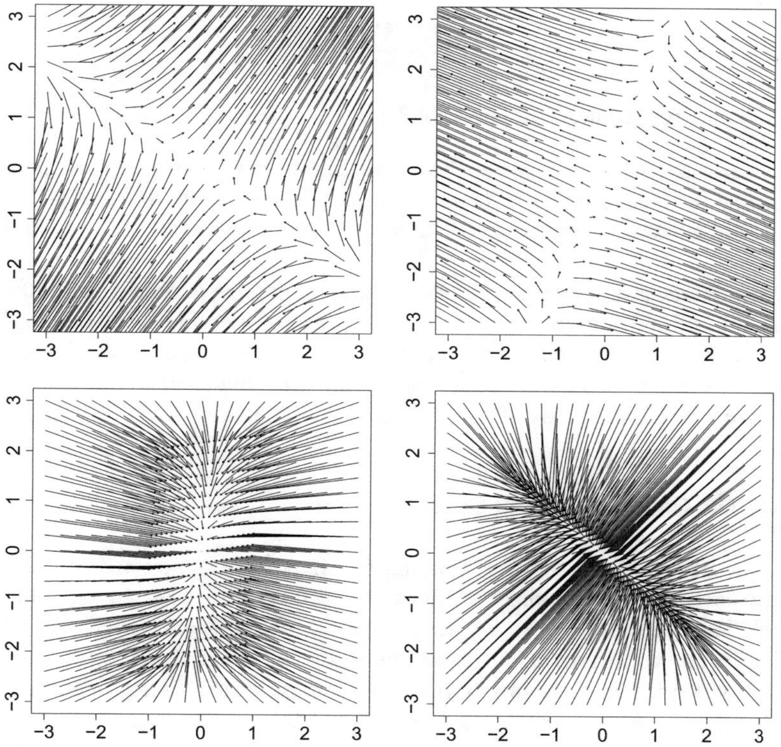

Fig. 5.3: Gaussian example: vector fields depicting the optimal maps $x \mapsto \mathbf{t}_{\bar{\mu}}^{\mu^i}(x)$ from the Fréchet mean $\bar{\mu}$ of Fig. 5.2 to the four measures $\{\mu^i\}$ of Fig. 5.1. The order corresponds to that of Fig. 5.1

5.4.2 *Compatible Measures*

We next discuss the behaviour of the algorithm when the measures are compatible. Recall that a collection $\mathscr{C} \subseteq \mathscr{W}_2(\mathscr{X})$ is *compatible* if for all $\gamma, \rho, \mu \in \mathscr{C}$, $\mathbf{t}_\mu^\nu \circ \mathbf{t}_\gamma^\mu = \mathbf{t}_\gamma^\nu$ in $L_2(\gamma)$ (Definition 2.3.1). Boissard et al. [28] showed that when this condition holds, the Fréchet mean of $(\mu^1, \ldots, \mu^N)$ can be found by simple computations involving the *iterated barycentre*. We again denote by γ_0 the initial point of Algorithm 1, which can be any absolutely continuous measure.

Lemma 5.4.2 (Compatibility and Convergence) *If $\gamma_0 \cup \{\mu^i\}$ is compatible, then Algorithm 1 converges to the Fréchet mean of (μ^i) after a single step.*

Proof. By definition, the next iterate is

$$\gamma_1 = \left[\frac{1}{N}\sum_{i=1}^{N} \mathbf{t}_{\gamma_0}^{\mu^i}\right] \# \gamma_0,$$

which is the Fréchet mean by Theorem 3.1.9.

In this case, Algorithm 1 requires the calculation of N pairwise optimal maps, and this can be reduced to $N-1$ if the initial point is chosen to be μ^1. This is the same computational complexity as the calculation of the iterated barycentre proposed in [28].

When the measures have a common copula, finding the optimal maps reduces to finding the optimal maps between the one-dimensional marginals (see Lemma 2.3.3) and this can be done using quantile functions as described in Sect. 1.5. The marginal Fréchet means are then plugged into the common copula to yield the joint Fréchet mean. We next illustrate Algorithm 1 in three such scenarios.

5.4.2.1 The One-Dimensional Case

When the measures are supported on the real line, there is no need to use the algorithm since the Fréchet mean admits a closed-form expression in terms of quantile functions (see Sect. 3.1.4). We nevertheless discuss this case briefly because we build upon this construction in subsequent examples. Given that $\mathbf{t}_\mu^\nu = F_\nu^{-1} \circ F_\mu$, we may apply Algorithm 1 starting from one of these measures (or any other measure). Figure 5.4 plots $N = 4$ univariate densities and the Fréchet mean yielded by the algorithm in two different scenarios. At the left, the densities were generated as

$$f^i(x) = \frac{1}{2}\phi\left(\frac{x-m_1^i}{\sigma_1^i}\right) + \frac{1}{2}\phi\left(\frac{x-m_2^i}{\sigma_2^i}\right), \tag{5.5}$$

with ϕ the standard normal density, and the parameters generated independently as

$$m_1^i \sim U[-13,-3], \quad m_2^i \sim U[3,13], \quad \sigma_1^i, \sigma_2^i \sim Gamma(4,4).$$

At the right of Fig. 5.4, we used a mixture of a shifted gamma and a Gaussian:

$$f^i(x) = \frac{3}{5}\frac{\beta_i^3}{\Gamma(3)}(x-m_3^i)^2 e^{-\beta_i(x-m_3^i)} + \frac{2}{5}\phi(x-m_4^i), \tag{5.6}$$

with

$$\beta^i \sim Gamma(4,1), \quad m_3^i \sim U[1,4], \quad m_4^i \sim U[-4,-1].$$

The resulting Fréchet mean density for both settings is shown in thick light blue, and can be seen to capture the bimodal nature of the data. Even though the Fréchet mean of Gaussian mixtures is not a Gaussian mixture itself, it is approximately so, provided that the peaks are separated enough. Figure 5.5 shows the optimal maps pushing the Fréchet mean $\bar{\mu}$ to the measures $\mu^1, \ldots, \mu^N$ in each case. If one ignores the "middle part" of the x axis, the maps appear (approximately) affine for small values of x and for large values of x, indicating how the peaks are shifted. In the middle region, the maps need to "bridge the gap" between the different slopes and intercepts of these affine maps.

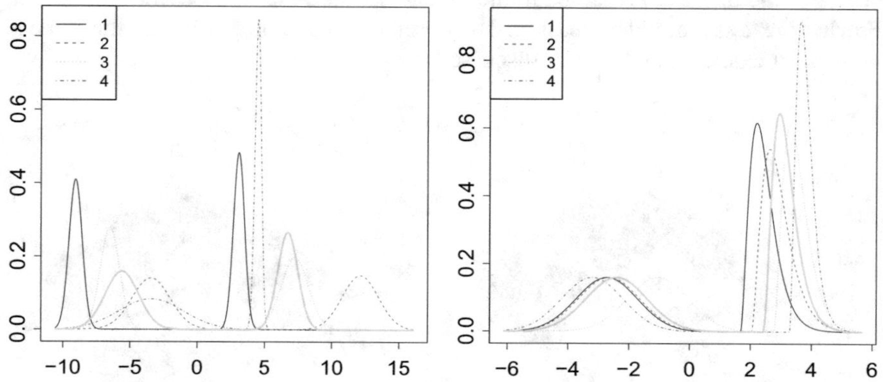

Fig. 5.4: Densities of a bimodal Gaussian mixture (left) and a mixture of a Gaussian with a gamma (right), with the Fréchet mean density in light blue

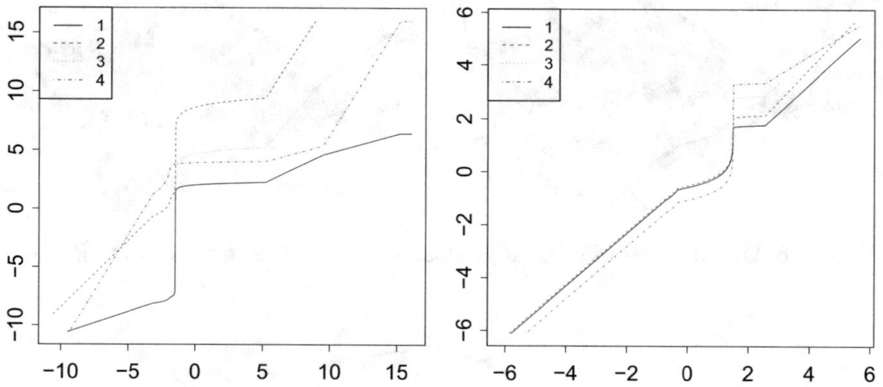

Fig. 5.5: Optimal maps $\mathbf{t}_{\bar{\mu}}^{\mu^i}$ from the Fréchet mean $\bar{\mu}$ to the four measures $\{\mu^i\}$ in Fig. 5.4. The left plot corresponds to the bimodal Gaussian mixture, and the right plot to the Gaussian/gamma mixture

5.4.2.2 Independence

We next take measures μ^i on $\mathbb{R}^2$, having independent marginal densities f_X^i as in (5.5), and f_Y^i as in (5.6). Figure 5.6 shows the density plot of $N = 4$ such measures, constructed as the product of the measures from Fig. 5.4. One can distinguish the independence by the "parallel" structure of the figures: for every pair (y_1, y_2), the ratio $g(x,y_1)/g(x,y_2)$ does not depend on x (and vice versa, interchanging x and y). Figure 5.7 plots the density of the resulting Fréchet mean. We observe that the Fréchet mean captures the four peaks and their location. Furthermore, the parallel nature of the figure is preserved in the Fréchet mean. Indeed, by Lemma 3.1.11 the Fréchet mean is a product measure. The optimal maps, in Fig. 5.10, are the same as in the next example, and will be discussed there.

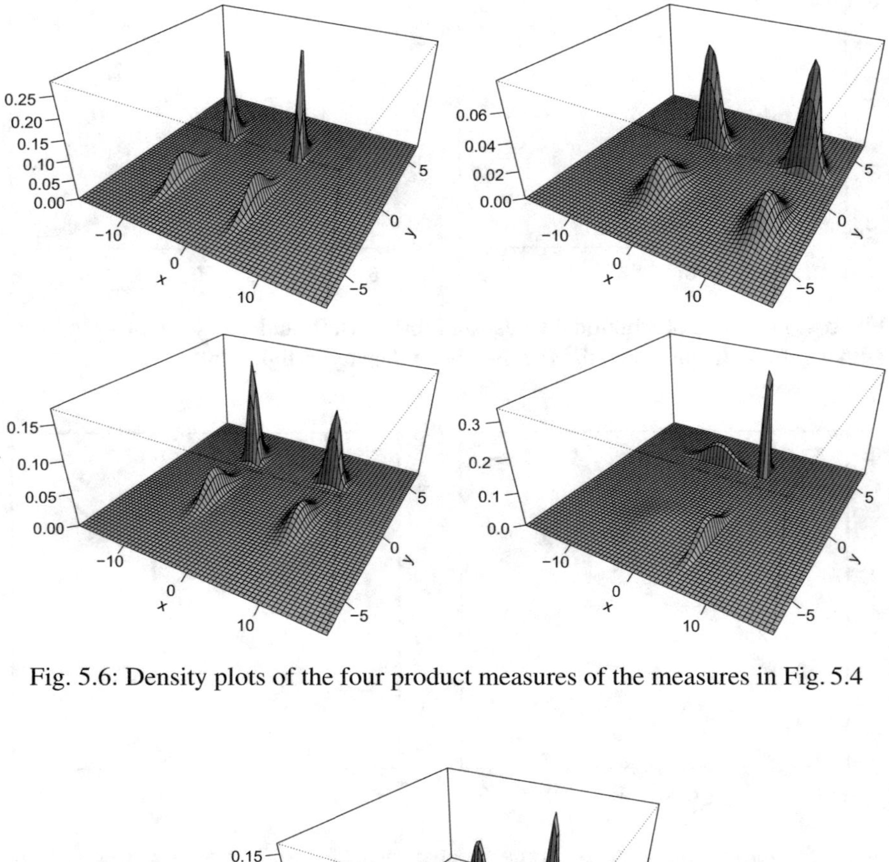

Fig. 5.6: Density plots of the four product measures of the measures in Fig. 5.4

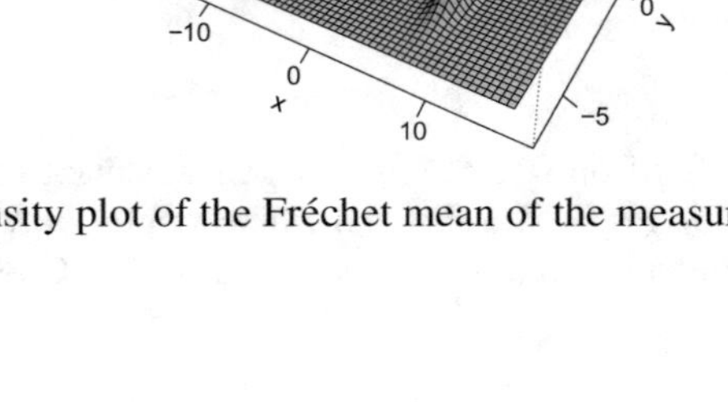

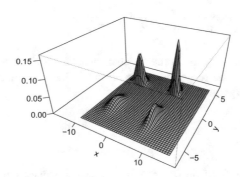

Fig. 5.7: Density plot of the Fréchet mean of the measures in Fig. 5.6

5.4.2.3 Common Copula

Let μ^i be a measure on $\mathbb{R}^2$ with density

$$g^i(x,y) = c(F_X^i(x), F_Y^i(y)) f_X^i(x) f_Y^i(y),$$

where f_X^i and f_Y^i are random densities on the real line with distribution functions F_X^i and F_Y^i, and c is a copula density. Figure 5.8 shows the density plot of $N = 4$ such measures, with f_X^i generated as in (5.5), f_Y^i as in (5.6), and c is the Frank(-8) copula density, while Fig. 5.9 plots the density of the Fréchet mean obtained. (For ease of comparison we use the same realisations of the densities that appear in Fig. 5.4.) The Fréchet mean can be seen to preserve the shape of the density, having four clearly distinguished peaks. Figure 5.10, depicting the resulting optimal maps, allows for a clearer interpretation: for instance, the leftmost plot (in black) shows more clearly that the map splits the mass around $x = -2$ to a much wider interval; and conversely a very large amount mass is sent to $x \approx 2$. This rather extreme behaviour matches the peak of the density of μ^1 located at $x = 2$.

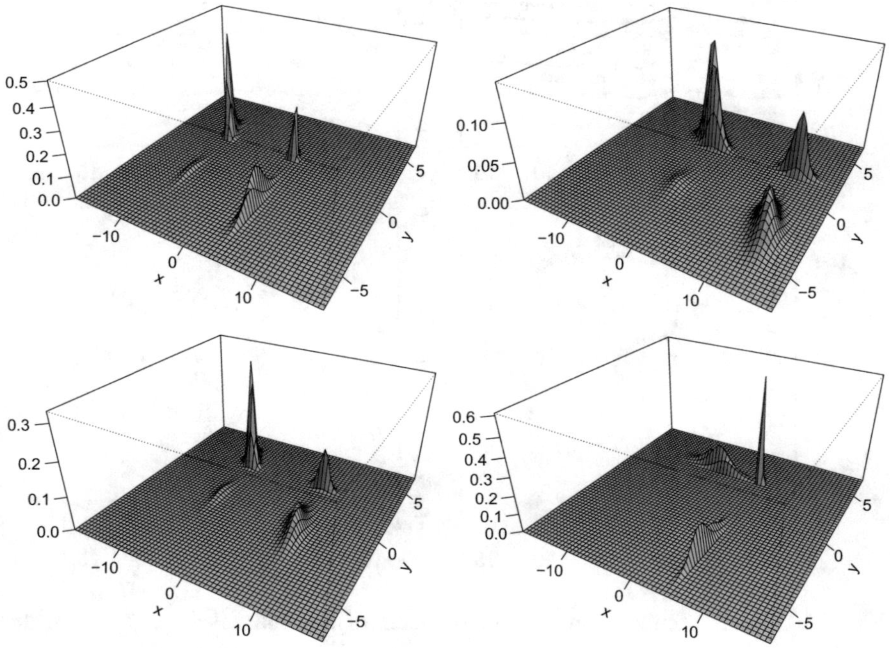

Fig. 5.8: Density plots of four measures in $\mathbb{R}^2$ with Frank copula of parameter -8

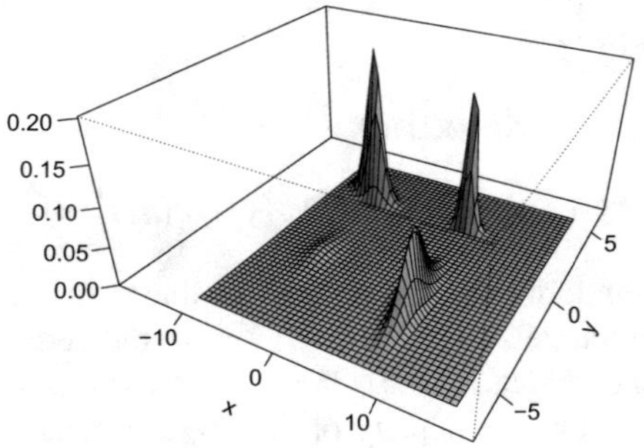

Fig. 5.9: Density plot of the Fréchet mean of the measures in Fig. 5.8

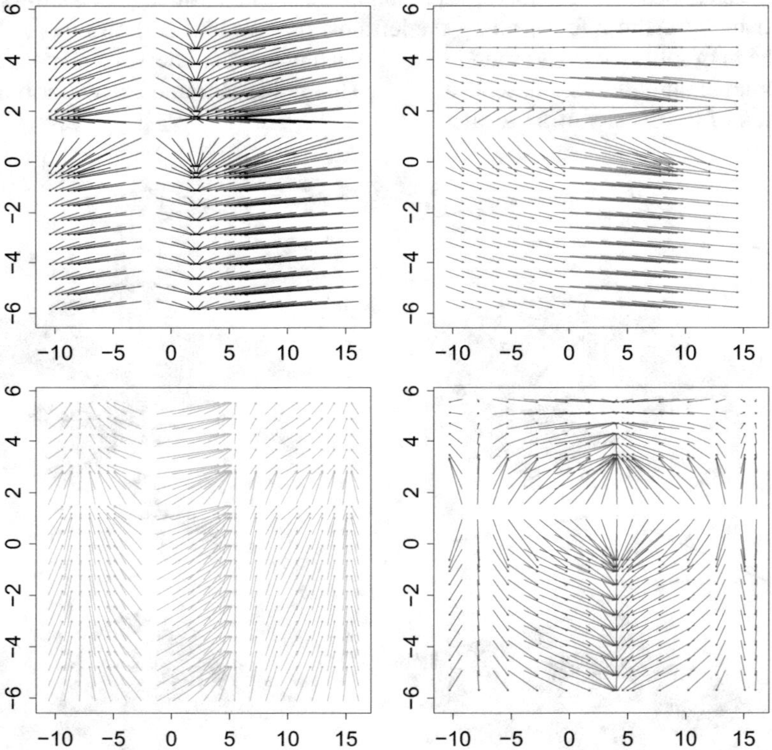

Fig. 5.10: Frank copula example: vector fields of the optimal maps $\mathbf{t}_{\bar{\mu}}^{\mu^i}$ from the Fréchet mean $\bar{\mu}$ of Fig. 5.9 to the four measures $\{\mu^i\}$ of Fig. 5.8. The colours match those of Fig. 5.4

5.4.3 Partially Gaussian Trivariate Measures

We now apply Algorithm 1 in a situation that entangles two of the previous settings. Let U be a fixed 3×3 real orthogonal matrix with columns U_1, U_2, U_3 and let μ^i have density

$$g^i(y_1,y_2,y_3) = g^i(y) = f^i(U_3^t y)\frac{1}{2\pi\sqrt{\det S^i}}\exp\left[-\frac{(U_1^t y, U_2^t y)(S^i)^{-1}\binom{U_1^t y}{U_2^t y}}{2}\right],$$

with f^i bounded density on the real line and $S^i \in \mathbb{R}^{2\times 2}$ positive definite. We simulated $N = 4$ such densities with f^i as in (5.5) and $S^i \sim \text{Wishart}(I_2, 2)$. We apply Algorithm 1 to this collection of measures and find their Fréchet mean (see the end of this subsection for precise details on how the optimal maps were calculated). Figure 5.11 shows level set of the resulting densities for some specific values. The bimodal nature of f^i implies that for most values of a, $\{x : f^i(x) = a\}$ has four elements. Hence, the level sets in the figures are unions of four separate parts, with each peak of f^i contributing two parts that form together the boundary of an ellipsoid in $\mathbb{R}^3$ (see Fig. 5.12). The principal axes of these ellipsoids and their position in $\mathbb{R}^3$ differ between the measures, but the Fréchet mean can be viewed as an average of those in some sense.

In terms of orientation (principal axes) of the ellipsoids, the Fréchet mean is most similar to μ^1 and μ^2, whose orientations are similar to one another.

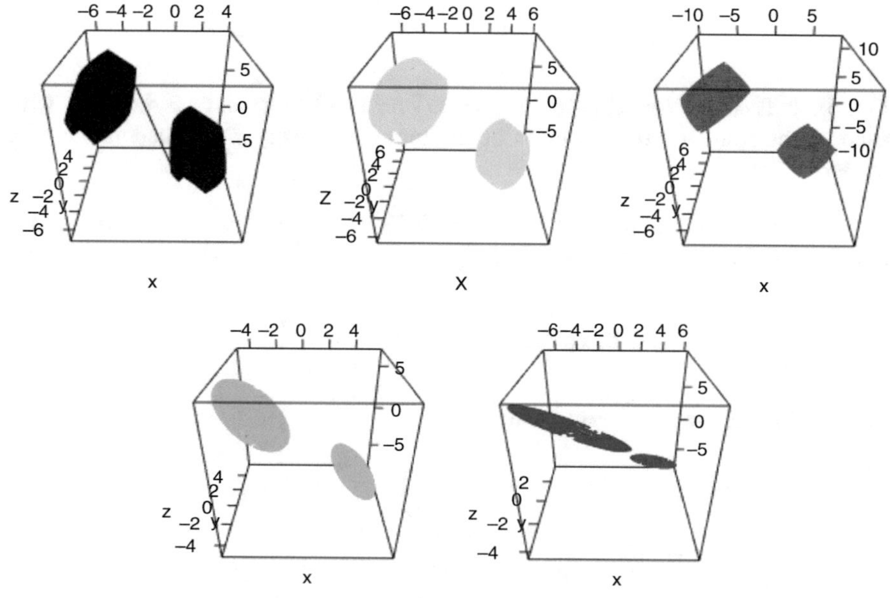

Fig. 5.11: The set $\{v \in \mathbb{R}^3 : g^i(v) = 0.0003\}$ for $i = 1$ (black), the Fréchet mean (light blue), $i = 2, 3, 4$ in red, green, and dark blue, respectively

Let us now see how the optimal maps are calculated. If $Y=(y_1,y_2,y_3)\sim\mu^i$, then the random vector $(x_1,x_2,x_3)=X=U^{-1}Y$ has joint density

$$f^i(x_3)\exp\left[-\frac{(x_1,x_2)(\Sigma^i)^{-1}\binom{x_1}{x_2}}{2}\right]\frac{1}{2\pi\sqrt{\det\Sigma^i}},$$

so the probability law of X is $\rho^i\otimes\nu^i$ with ρ^i centred Gaussian with covariance matrix Σ^i and ν^i having density f^i on $\mathbb{R}$. By Lemma 3.1.11, the Fréchet mean of $(U^{-1}\#\mu^i)$ is the product measure of that of (ρ^i) and that of (ν^i); by Lemma 3.1.12, the Fréchet mean of (μ^i) is therefore

$$U\#(\mathscr{N}(0,\Sigma)\otimes f),\qquad f=F',\quad F^{-1}(q)=\frac{1}{N}\sum_{i=1}^N F_i^{-1}(q),\quad F_i(x)=\int_{-\infty}^x f^i(s)\,\mathrm{d}s,$$

where Σ is the Fréchet–Wasserstein mean of $\Sigma_1,\dots,\Sigma_N$.

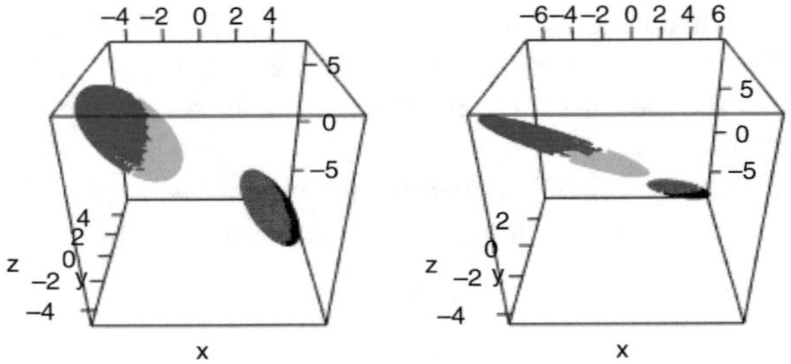

Fig. 5.12: The set $\{v\in\mathbb{R}^3: g^i(v)=0.0003\}$ for $i=3$ (left) and $i=4$ (right), with each of the four different inverses of the bimodal density f^i corresponding to a colour

Starting at an initial point $\gamma_0=U\#(\mathscr{N}(0,\Sigma_0)\otimes\nu_0)$, with ν_0 having continuous distribution F_{ν_0}, the optimal maps are $U\circ\mathbf{t}_0^i\circ U^{-1}=\nabla(\varphi_0^i\circ U^{-1})$ with

$$\mathbf{t}_0^i(x_1,x_2,x_3)=\begin{pmatrix}\mathbf{t}_{\Sigma_0}^{\Sigma^j}(x_1,x_2)\\ F_j^{-1}\circ F_{\nu_0}(x_3)\end{pmatrix}$$

the gradients of the convex function

$$\varphi_0^i(x_1,x_2,x_3)=(x_1,x_2)\mathbf{t}_{\gamma_0}^{\Sigma^i}\begin{pmatrix}x_1\\x_2\end{pmatrix}+\int_0^{x_3}F_j^{-1}(F_{\nu_0}(s))\,\mathrm{d}s,$$

where we identify $\mathbf{t}_{\gamma_0}^{\Sigma^i}$ with the positive definite matrix $(\Sigma^i)^{1/2}[(\Sigma^i)^{1/2}\Sigma_0(\Sigma^i)^{1/2}]^{-1/2}(\Sigma^i)^{1/2}$ that pushes forward $\mathscr{N}(0,\Sigma_0)$ to $\mathscr{N}(0,\Sigma^i)$. Due to the one-dimensionality, the algorithm finds the third component of the rotated measures after one step, but the convergence of the Gaussian component requires further iterations.

5.5 Population Version of Algorithm 1

Let $\Lambda \in \mathscr{W}_2(\mathbb{R}^d)$ be a random measure with finite Fréchet functional. The population version of (5.1) is

$$q = \mathbb{P}(\Lambda \text{ absolutely continuous with density bounded by } M) > 0 \quad \text{for some } M < \infty, \tag{5.7}$$

which we assume henceforth. This condition is satisfied if and only if

$$\mathbb{P}(\Lambda \text{ absolutely continuous with bounded density}) > 0.$$

These probabilities are well-defined because the set

$$\mathscr{W}_2(\mathbb{R}^d; M) = \{\mu \in \mathscr{W}_2(\mathbb{R}^d) : \mu \text{ absolutely continuous with density bounded by } M\}$$

is weakly closed (see the paragraph before Proposition 5.3.6), hence a Borel set of $\mathscr{W}_2(\mathbb{R}^d)$.

In light of Theorem 3.2.13, we can define a population version of Algorithm 1 with the iteration function

$$\mathscr{A}(\gamma) = \mathbb{E}\mathbf{t}_\gamma^\Lambda, \qquad \gamma \in \mathscr{W}_2(\mathbb{R}^d) \text{ absolutely continuous.}$$

The (Bochner) expectation is well-defined in $\mathscr{L}_2(\gamma)$ because the random map $\mathbf{t}_\gamma^\Lambda$ is measurable (Lemma 2.4.6). Since $\mathscr{L}_2(\gamma)$ is a Hilbert space, the law of large numbers applies there, and results for the empirical version carry over to the population version by means of approximations. In particular:

Lemma 5.5.1 *Any descent iterate γ has density bounded by $q^{-d}M$, where q and M are as in* (5.7).

Proof. The result is true in the empirical case, by Proposition 5.3.6. The key point (observed by Pass [102, Subsection 3.3]) is that the number of measures does not appear in the bound $q^{-d}M$.

Let $\Lambda_1, \dots$ be a sample from Λ and let q_n be the proportion of measures in $(\Lambda_1, \dots, \Lambda_n)$ that have density bounded by M. Then both $n^{-1}\sum_{i=1}^n \mathbf{t}_\gamma^{\Lambda_i} \to \mathbb{E}\mathbf{t}_\gamma^\Lambda$ and $q_n \to q$ almost surely by the law of large numbers. Pick any ω in the probability space for which this happens and notice that (invoking Lemma 2.4.5)

$$\mathscr{A}(\gamma) = \left[\lim_{n\to\infty} \frac{1}{n}\sum_{i=1}^n \mathbf{t}_\gamma^{\Lambda_i}\right]\#\gamma = \lim_{n\to\infty}\left[\frac{1}{n}\sum_{i=1}^n \mathbf{t}_\gamma^{\Lambda_i}\right]\#\gamma.$$

Let λ_n denote the measure in the last limit. By Proposition 5.3.6, its density is bounded by $q_n^{-d}M \to q^{-d}M$ almost surely, so for any $C > q^{-d}M$ and n large, λ_n has density bounded by C. By the portmanteau Lemma 1.7.1, so does $\lim \lambda_n = [\mathbb{E}\mathbf{t}_\gamma^\Lambda]\#\gamma$. Now let $C \searrow q^{-d}M$.

Though it follows that every Karcher mean of Λ has a bounded density, we cannot yet conclude that the same bound holds for the Fréchet mean, because we need an a-priori knowledge that the latter is absolutely continuous. This again can be achieved by approximations:

Theorem 5.5.2 (Bounded Density for Population Fréchet Mean) *Let $\Lambda \in \mathscr{W}_2(\mathbb{R}^d)$ be a random measure with finite Fréchet functional. If Λ has a bounded density with positive probability, then the Fréchet mean of Λ is absolutely continuous with a bounded density.*

Proof. Let q and M be as in (5.7), $\Lambda_1,\dots$ be a sample from Λ, and q_n the proportion of $(\Lambda_i)_{i\le n}$ with density bounded by M. The empirical Fréchet mean λ_n of the sample $(\Lambda_1,\dots,\Lambda_n)$ has a density bounded by $q_n^{-d}M$. The Fréchet mean λ of Λ is unique by Proposition 3.2.7, and consequently $\lambda_n \to \lambda$ in $\mathscr{W}_2(\mathbb{R}^d)$ by the law of large numbers (Corollary 3.2.10). For any $C > \limsup q_n^{-d}M$, the density of λ is bounded by C by the portmanteau Lemma 1.7.1, and the limsup is $q^{-d}M$ almost surely. Thus, the density is bounded by $q^{-d}M$.

In the same way, one shows the population version of Theorem 3.1.9:

Theorem 5.5.3 (Fréchet Mean of Compatible Measures) *Let $\Lambda : (\Omega,\mathscr{F},\mathbb{P}) \to \mathscr{W}_2(\mathbb{R}^d)$ be a random measure with finite Fréchet functional, and suppose that with positive probability Λ is absolutely continuous and has bounded density. If the collection $\{\gamma\}\cup\Lambda(\Omega)$ is compatible and γ is absolutely continuous, then $[\mathbb{E}\mathbf{t}_\gamma^\Lambda]\#\gamma$ is the Fréchet mean of Λ.*

It is of course sufficient that $\{\gamma\}\cup\Lambda(\Omega\setminus\mathscr{N})$ be compatible for some null set $\mathscr{N}\subset\Omega$.

5.6 Bibliographical Notes

The algorithm outlined in this chapter was suggested independently in this steepest descent form by Zemel and Panaretos [134] and in the form a fixed point equation iteration by Álvarez-Esteban et al. [9]. These two papers provide different alternative proofs of Theorem 5.3.1. The exposition here is based on [134]. Although longer and more technical than the one in [9], the formalism in [134] is amenable to directly treating the optimal maps (Theorem 5.3.3) and the multicouplings (Corollary 5.3.4). On the flip side, it is noteworthy that the proof of the Gaussian case (Theorem 5.4.1) given in [9] is more explicit and quantitative; for instance, it shows the additional property that the traces of the matrix iterates are monotonically increasing.

Developing numerical schemes for computing Fréchet means in $\mathscr{W}_2(\mathbb{R}^d)$ is a very active area of research, and readers are referred to the recent monograph of Peyré and Cuturi [103, Section 9.2] for a survey.

In recent work, Backhoff-Varaguas et al. [15] propose a stochastic steepest descent for finding Karcher means of a population Fréchet functional associated with a random measure Λ. At iterate j, one replaces γ_j by

$$[t_j \mathbf{t}_{\gamma_j}^{\mu_j} + (1 - t_j)\mathbf{i}]\#\gamma_j, \qquad \mu_j \sim \Lambda.$$

The analogue of Theorem 5.3.1 holds under further conditions.

References

1. B. Afsari, R. Tron, R. Vidal, On the convergence of gradient descent for finding the Riemannian center of mass. SIAM J. Contr. Opt. **51**(3), 2230–2260 (2013)
2. M. Agueh, G. Carlier, Barycenters in the Wasserstein space. Soc. Indus. Appl. Math. **43**(2), 904–924 (2011)
3. M. Agueh, G. Carlier, Vers un théorème de la limite centrale dans l'espace de Wasserstein? C.R. Math. **355**(7), 812–818 (2017)
4. A. Ahidar-Coutrix, T. Le Gouic, Q. Paris, Convergence rates for empirical barycenters in metric spaces: curvature, convexity and extendable geodesics. Probab. Theory Relat. Fields (2019). https://link.springer.com/article/10.1007%2Fs00440-019-00950-0
5. M. Ajtai, J. Komlós, G. Tusnády, On optimal matchings. Combinatorica **4**(4), 259–264 (1984)
6. G. Alberti, L. Ambrosio, A geometrical approach to monotone functions in $\mathbb{R}^n$. Math. Z. **230**(2), 259–316 (1999)
7. S. Alesker, S. Dar, V. Milman, A remarkable measure preserving diffeomorphism between two convex bodies in $\mathbb{R}^n$. Geometriae Dedicata **74**(2), 201–212 (1999)
8. P.C. Álvarez-Esteban, E. del Barrio, J.A. Cuesta-Albertos, C. Matrán, Uniqueness and approximate computation of optimal incomplete transportation plans. Ann. Inst. Henri Poincaré Probab. Stat. **47**(2), 358–375 (2011)
9. P.C. Álvarez-Esteban, E. del Barrio, J.A. Cuesta-Albertos, C. Matrán, A fixed-point approach to barycenters in Wasserstein space. J. Math. Anal. Appl. **441**(2), 744–762 (2016)
10. L. Ambrosio, N. Gigli, A user's guide to optimal transport, in *Modelling and Optimisation of Flows on Networks* (Springer, Berlin, 2013), pp. 1–155
11. L. Ambrosio, A. Pratelli, Existence and stability results in the L^1-theory of optimal transportation, in *Optimal Transportation and Applications* (Springer, Berlin, 2003), pp. 123–160

V. M. Panaretos, Y. Zemel, *An Invitation to Statistics in Wasserstein Space*,
SpringerBriefs in Probability and Mathematical Statistics,
https://doi.org/10.1007/978-3-030-38438-8

12. L. Ambrosio, N. Gigli, G. Savaré, *Gradient Flows in Metric Spaces and in the Space of Probability Measures*. Lectures in Mathematics. ETH Zürich, 2nd edn. (Springer, Berlin, 2008)
13. Y. Amit, U. Grenander, M. Piccioni, Structural image restoration through deformable templates. J. Amer. Stat. Assoc. **86**(414), 376–387 (1991)
14. E. Anderes, S. Borgwardt, J. Miller, Discrete Wasserstein barycenters: optimal transport for discrete data. Math. Meth. Oper. Res. **84**(2), 1–21 (2016)
15. J. Backhoff-Veraguas, J. Fontbona, G. Rios, F. Tobar, Bayesian learning with Wasserstein barycenters (2018). arXiv:1805.10833
16. A. Bandeira, P. Rigollet, J. Weed, Optimal rates of estimation for multi-reference alignment (2017). arXiv:1702.08546
17. M. Beiglböck, W. Schachermayer, Duality for Borel measurable cost functions. Trans. Amer. Math. Soc. **363**(8), 4203–4224 (2011)
18. M. Beiglböck, M. Goldstern, G. Maresch, W. Schachermayer, Optimal and better transport plans. J. Func. Anal. **256**(6), 1907–1927 (2009)
19. D.P. Bertsekas, *Nonlinear Programming* (Athena Scientific, Belmont, 1999)
20. R. Bhatia, *Positive Definite Matrices* (Princeton University Press, Princeton, 2009)
21. P.J. Bickel, D.A. Freedman, Some asymptotic theory for the bootstrap. Ann. Stat. **9**(6), 1196–1217 (1981)
22. J. Bigot, T. Klein, Characterization of barycenters in the Wasserstein space by averaging optimal transport maps. ESAIM: Probab. Stat. **22**, 35–57 (2018)
23. J. Bigot, R. Gouet, T. Klein, A. López, Upper and lower risk bounds for estimating the Wasserstein barycenter of random measures on the real line. Electron. J. Stat. **12**(2), 2253–2289 (2018)
24. P. Billingsley, *Convergence of Probability Measures*, 2nd edn. (Wiley, New York, 1999)
25. S. Bobkov, M. Ledoux, *One-Dimensional Empirical Measures, Order Statistics and Kantorovich Transport Distances*, vol. 261, no. 1259 (Memoirs of the American Mathematical Society, Providence, 2019). https://doi.org/10.1090/memo/1259
26. V.I. Bogachev, A.V. Kolesnikov, The Monge–Kantorovich problem: achievements, connections, and perspectives. Russ. Math. Surv. **67**(5), 785–890 (2012)
27. E. Boissard, T. Le Gouic, On the mean speed of convergence of empirical and occupation measures in Wasserstein distance. Ann. Inst. H. Poincaré. Probab. Stat. **50**(2), 539–563 (2014)
28. E. Boissard, T. Le Gouic, J.-M. Loubes, Distribution's template estimate with Wasserstein metrics. Bernoulli **21**(2), 740–759 (2015)
29. F. Bolley, A. Guillin, C. Villani, Quantitative concentration inequalities for empirical measures on non-compact spaces. Prob. Theory Rel. Fields **137**, 541–593 (2007)
30. N. Bonneel, J. Rabin, G. Peyré, H. Pfister, Sliced and radon Wasserstein barycenters of measures. J. Math. Imag. Vis. **51**(1), 22–45 (2015)

31. Y. Brenier, Polar factorization and monotone rearrangement of vector-valued functions. Commun. Pure Appl. Math. **44**(4), 375–417 (1991)
32. L.A. Caffarelli, The regularity of mappings with a convex potential. J. Amer. Math. Soc. **5**(1), 99–104 (1992)
33. G. Carlier, I. Ekeland, Matching for teams. Econ. Theory **42**(2), 397–418 (2010)
34. A. Chakraborty, V.M. Panaretos, Functional registration and local variations: Identifiability, rank, and tuning (2017). arXiv:1702.03556
35. V. Chernozhukov, A. Galichon, M. Hallin, M. Henry, Monge–Kantorovich depth, quantiles, ranks and signs. Ann. Stat. **45**(1), 223–256 (2017)
36. P. Clément, W. Desch, An elementary proof of the triangle inequality for the Wasserstein metric. Proc. Amer. Math. Soc. **136**(1), 333–339 (2008)
37. J.A. Cuesta-Albertos, C. Matrán, Notes on the Wasserstein metric in Hilbert spaces. Ann. Probab. **17**(3), 1264–1276 (1989)
38. J.A. Cuesta-Albertos, L. Rüschendorf, A. Tuero-Diaz, Optimal coupling of multivariate distributions and stochastic processes. J. Multivar. Anal. **46**(2), 335–361 (1993)
39. J.A. Cuesta-Albertos, C. Matrán-Bea, A. Tuero-Diaz, On lower bounds for the L_2-Wasserstein metric in a Hilbert space. J. Theor. Probab. **9**(2), 263–283 (1996)
40. J.A. Cuesta-Albertos, C. Matrán, A. Tuero-Diaz, Optimal transportation plans and convergence in distribution. J. Multivar. Anal. **60**(1), 72–83 (1997)
41. D.J. Daley, D. Vere-Jones, *An Introduction to the Theory of Point Processes: Volume II: General Theory and Structure* (Springer, Berlin, 2007)
42. E. del Barrio, J.-M. Loubes, Central limit theorems for empirical transportation cost in general dimension. Ann. Probab. **47**(2), 926–951 (2019). https://projecteuclid.org/euclid.aop/1551171641
43. E. del Barrio, E. Giné, C. Matrán, Central limit theorems for the Wasserstein distance between the empirical and the true distributions. Ann. Probab. **27**(2), 1009–1071 (1999)
44. D. Dowson, B. Landau, The Fréchet distance between multivariate normal distributions. J. Multivar. Anal. **12**(3), 450–455 (1982)
45. I.L. Dryden, K.V. Mardia, *Statistical Shape Analysis*, vol. 4 (Wiley, Chichester, 1998)
46. R.M. Dudley, The speed of mean Glivenko–Cantelli convergence. Ann. Math. Stat. **40**(1), 40–50 (1969)
47. R.M. Dudley, *Real Analysis and Probability*, vol. 74 (Cambridge University Press, Cambridge, 2002)
48. N. Dunford, J.T. Schwartz, W.G. Bade, R.G. Bartle, *Linear Operators* (Wiley-Interscience, New York, 1971)
49. R. Durrett, *Probability: Theory and Examples* (Cambridge University Press, Cambridge, 2010)
50. J. Edmonds, R.M. Karp, Theoretical improvements in algorithmic efficiency for network flow problems. J. ACM **19**(2), 248–264 (1972)

51. F. Ferraty, P. Vieu, *Nonparametric Functional Data Analysis: Theory and Practice* (Springer, Beriln, 2006)
52. A. Figalli, *The Monge–Ampère Equation and Its Applications* (European Mathematical Society, Zürich, 2017)
53. J. Fontbona, H. Guérin, S. Méléard, Measurability of optimal transportation and strong coupling of martingale measures. Electron. Commun. Probab. **15**, 124–133 (2010)
54. N. Fournier, A. Guillin, On the rate of convergence in Wasserstein distance of the empirical measure. Probab. Theory Rel. Fields **162**(3–4), 707–738 (2015)
55. M. Fréchet, Les éléments aléatoires de nature quelconque dans un espace distancié. Ann. Inst. Henri Poincaré **10**(4), 215–310 (1948)
56. M. Fréchet, Sur la distance de deux lois de probabilité. C.R. Hebd. Seances Acad. Sci. **244**(6), 689–692 (1957)
57. D.H. Fremlin, *Measure Theory, Vol. 4: Topological Measure Theory* (Torres Fremlin, Colchester, 2003)
58. B. Galasso, Y. Zemel, M. de Carvalho, Bayesian semiparametric modelling of phase-varying point processes (2018). arXiv:1812.09607
59. W. Gangbo, R.J. McCann, The geometry of optimal transportation. Acta Math. **177**(2), 113–161 (1996)
60. W. Gangbo, A. Święch, Optimal maps for the multidimensional Monge–Kantorovich problem. Comm. Pure Appl. Math. **51**(1), 23–45 (1998)
61. T. Gasser, A. Kneip, Searching for structure in curve samples. J. Amer. Statist. Assoc. **90**(432), 1179–1188 (1995)
62. M. Gelbrich, On a formula for the L_2-Wasserstein metric between measures on Euclidean and Hilbert spaces. Math. Nachr. **147**(1), 185–203 (1990)
63. D. Gervini, T. Gasser, Self-modelling warping functions. J. Roy. Stat. Soc.: Ser. B **66**(4), 959–971 (2004)
64. D. Gervini, T. Gasser, Nonparametric maximum likelihood estimation of the structural mean of a sample of curves. Biometrika **92**(4), 801–820 (2005)
65. C.R. Givens, R.M. Shortt, A class of Wasserstein metrics for probability distributions. Mich. Math. J. **31**(2), 231–240 (1984)
66. J.C. Gower, Generalized Procrustes analysis. Psychometrika **40**(1), 33–51 (1975)
67. D. Groisser, On the convergence of some Procrustean averaging algorithms. Stochastics: An Inter. J. Probab. Stoch. Process. **77**(1), 31–60 (2005)
68. H. Heinich, J.-C. Lootgieter, Convergence des fonctions monotones. C. R. Acad. Sci. ser. 1 Mathé **322**(9), 869–874 (1996)
69. T. Hildebrandt, Integration in abstract spaces. Bull. Amer. Math. Soc. **59**(2), 111–139 (1953)
70. L. Horváth, P. Kokoszka, *Inference for Functional Data with Applications*, vol. 200 (Springer, Berlin, 2012)
71. T. Hsing, R. Eubank, *Theoretical Foundations of Functional Data Analysis, with an Introduction to Linear Operators* (Wiley, Hoboken, 2015)
72. G.M. James, Curve alignment by moments. Ann. Appl. Stat. **1**(2), 480–501 (2007)

73. H.E. Jones, N. Bayley, The Berkeley growth study. Child Dev. **12**(2), 167–173 (1941)
74. R. Jordan, D. Kinderlehrer, F. Otto, The variational formulation of the Fokker–Planck equation. SIAM J. Math. Anal. **29**(1), 1–17 (1998)
75. O. Kallenberg, *Random Measures*, 3rd edn. (Academic, Cambridge, 1983)
76. O. Kallenberg, *Foundations of Modern Probability* (Springer, Berlin, 1997)
77. L.V. Kantorovich, On the translocation of masses. (Dokl.) Acad. Sci. URSS **37**(3), 199–201 (1942)
78. H. Karcher, Riemannian center of mass and mollifier smoothing. Commun. Pure Appl. Math. **30**(5), 509–541 (1977)
79. A. Karr, *Point Processes and Their Statistical Inference*, vol. 7 (CRC Press, Boca Raton, 1991)
80. H.G. Kellerer, Duality theorems for marginal problems. Zeitschrift für Wahrscheinlichkeitstheorie und verwandte Gebiete **67**(4), 399–432 (1984)
81. B. Kloeckner, A generalization of Hausdorff dimension applied to Hilbert cubes and Wasserstein spaces. J. Topol. Anal. **4**(2), 203–235 (2012)
82. A. Kneip, J.O. Ramsay, Combining registration and fitting for functional models. J. Amer. Stat. Assoc. **103**(483), 1155–1165 (2008)
83. M. Knott, C.S. Smith, On the optimal mapping of distributions. J. Optim. Theory Appl. **43**(1), 39–49 (1984)
84. A. Kroshnin, V. Spokoiny, A. Suvorikova, Statistical inference for Bures–Wasserstein barycenters (2019). arXiv:1901.00226
85. H.W. Kuhn, The Hungarian method for the assignment problem. Nav. Res. Log. Quart. **2**, 83–97 (1955)
86. H. Le, Locating Fréchet means with application to shape spaces. Adv. Appl. Probab. **33**(2), 324–338 (2001)
87. T. Le Gouic, J.-M. Loubes, Existence and consistency of Wasserstein barycenters. Prob. Theory Relat. Fields **168**(3–4), 901–917 (2017)
88. E.L. Lehmann, A general concept of unbiasedness. Ann. Math. Stat. **22**(4), 587–592 (1951)
89. D.G. Luenberger, Y. Ye, *Linear and Nonlinear Programming* (Springer, Berlin, 2008)
90. J.S. Marron, J.O. Ramsay, L.M. Sangalli, A. Srivastava, Functional data analysis of amplitude and phase variation. Stat. Sci. **30**(4), 468–484 (2015)
91. V. Masarotto, V.M. Panaretos, Y. Zemel, Procrustes metrics on covariance operators and optimal transportation of Gaussian processes. Sankhyā A **81**, 172–213 (2019) (Invited Paper, Special Issue on Statistics on non-Euclidean Spaces and Manifolds)
92. D.M. Mason, A weighted approximation approach to the study of the empirical Wasserstein distance, in ed. by C. Houdré, D.M. Mason, P. Reynaud-Bouret, J. Rosiński, *High Dimensional Probability VII* (Birkhäuser, Basel, 2016), pp. 137–154
93. R.J. McCann, A convexity principle for interacting gases. Adv. Math. **128**(1), 153–179 (1997)

94. R.J. McCann, Exact solutions to the transportation problem on the line. Proc. R. Soc. London, Ser. A: Math. Phys. Eng. Sci. **455**(1984), 1341–1380 (1999)
95. G. Monge, Mémoire sur la théorie des déblais et des remblais. Histoire de l'Académie R. des Sci. de Paris **177**, 666–704 (1781)
96. J. Munkers, Algorithms for the assignment and transportation problems. J. Soc. Indust. Appl. Math. **5**, 32–38 (1957)
97. R.B. Nelsen, *An Introduction to Copulas* (Springer, Berlin, 2013)
98. I. Olkin, F. Pukelsheim, The distance between two random vectors with given dispersion matrices. Linear Algebra Appl. **48**, 257–263 (1982)
99. F. Otto, The geometry of dissipative evolution equations: the porous medium equation. Comm. Part. Differ. Equ. **26**, 101–174 (2001)
100. V.M. Panaretos, Y. Zemel, Amplitude and phase variation of point processes. Ann. Stat. **44**(2), 771–812 (2016)
101. V.M. Panaretos, Y. Zemel, Statistical aspects of Wasserstein distances. Annu. Rev. Stat. Appl. **6**, 405–431 (2019)
102. B. Pass, Optimal transportation with infinitely many marginals. J. Funct. Anal. **264**(4), 947–963 (2013)
103. G. Peyré, M. Cuturi, Computational optimal transport. Found. Trends Mach. Learn. **11**(5–6), 355–607 (2019). https://www.nowpublishers.com/article/Details/MAL-073
104. D. Pollard. *Convergence of Stochastic Processes* (Springer, Berlin, 2012)
105. A. Pratelli, On the sufficiency of c-cyclical monotonicity for optimality of transport plans. Mathe. Zeitschrift **258**(3), 677–690 (2008)
106. S.T. Rachev, The Monge–Kantorovich mass transference problem and its stochastic applications. Theory Probab. Appl. **29**(4), 647–676 (1985)
107. S.T. Rachev, L. Rüschendorf, *Mass Transportation Problems: Volume I: Theory, Volume II: Applications* (Springer, Berlin, 1998)
108. J. Ramsay, X. Li, Curve registration. J. Roy. Stat. Soc.: Ser. B (Stat. Methodol.) **60**(2), 351–363 (1998)
109. J.O. Ramsay, B.W. Silverman, *Applied Functional Data Analysis: Methods and Case Studies*, vol. 77 (Citeseer, 2002)
110. J.O. Ramsay, B.W. Silverman, *Functional Data Analysis*, 2nd edn. (Springer, Berlin, 2005)
111. J.O. Ramsay, H. Wickham, S. Graves, G. Hooker, *FDA: Functional Data Analysis*, R package version 2.4.8 (2018)
112. R.T. Rockafellar, Characterization of the subdifferentials of convex functions. Pac. J. Math. **17**(3), 497–510 (1966)
113. R.T. Rockafellar, *Convex Analysis* (Princeton University Press, Princeton, 1970)
114. C.A. Rogers, Covering a sphere with spheres. Mathematika **10**(2), 157–164 (1963)
115. B.B. Rønn, Nonparametric maximum likelihood estimation for shifted curves. J. Roy. Stat. Soc.: Ser. B (Stat. Methodol.) **63**(2), 243–259 (2001)
116. L. Rüschendorf, On c-optimal random variables. Stat. Probab. Lett. **27**(3), 267–270 (1996)

117. L. Rüschendorf, S.T. Rachev, A characterization of random variables with minimum L^2-distance. J. Multivar. Anal. **32**(1), 48–54 (1990)
118. H. Sakoe, S. Chiba, Dynamic programming algorithm optimization for spoken word recognition. IEEE Trans. Acoust. Speech Signal Process. **26**(1), 43–49 (1978)
119. F. Santambrogio, *Optimal Transport for Applied Mathematicians*, vol. 87 (Springer, Berlin, 2015)
120. W. Schachermayer, J. Teichmann, Characterization of optimal transport plans for the Monge–Kantorovich problem. Proc. Amer. Math. Soc. **137**(2), 519–529 (2009)
121. E.M. Stein, R. Shakarchi, *Real Analysis: Measure Theory, Integration & Hilbert Spaces*. (Princeton University Press, Princeton, 2005)
122. R. Tang, H.-G. Müller, Pairwise curve synchronization for functional data. Biometrika **95**(4), 875–889 (2008)
123. J.D. Tucker, W. Wu, A. Srivastava, Generative models for functional data using phase and amplitude separation. Comput. Stat. Data Anal. **61**, 50–66 (2013)
124. C. Villani, *Topics in Optimal Transportation* (American Mathematical Society, Providence, 2003)
125. C. Villani, *Optimal Transport: Old and New* (Springer, Berlin, 2008)
126. M.J. Wainwright, *High-Dimensional Statistics: A Non-Asymptotic Viewpoint* (Cambridge University Press, Cambridge, 2019)
127. J.-L. Wang, J.-M. Chiou, H.-G. Müller, Functional data analysis. Annu. Rev. Stat. Appl. **3**, 257–295 (2016)
128. K. Wang, T. Gasser, Alignment of curves by dynamic time warping. Ann. Stat. **25**(3), 1251–1276 (1997)
129. K. Wang, T. Gasser, Synchronizing sample curves nonparametrically. Ann. Stat. **27**, 439–460 (1999)
130. J. Weed, F. Bach, Sharp asymptotic and finite-sample rates of convergence of empirical measures in Wasserstein distance. Bernoulli **25**(4A), 2620–2648 (2019). https://projecteuclid.org/euclid.bj/1568362038
131. J. Wrobel, V. Zipunnikov, J. Schrack, J. Goldsmith, Registration for exponential family functional data. Biometrics **75**, 48–57 (2019)
132. S. Wu, H.-G. Müller, Z. Zhang, Functional data analysis for point processes with rare events. Stat. Sinica **23**(1), 1–23 (2013)
133. W. Wu, A. Srivastava, Analysis of spike train data: Alignment and comparisons using the extended Fisher–Rao metric. Electron. J. Stat. **8**, 1776–1785 (2014)
134. Y. Zemel, V.M. Panaretos, Fréchet means and Procrustes analysis in Wasserstein space. Bernoulli **25**(2), 932–976 (2019)
135. Y. Zemel, V.M. Panaretos, Supplement to "Fréchet means and Procrustes analysis in Wasserstein space" (2019)

Printed in the United States
By Bookmasters